Errata:
p. 103 Eq. 62
p. 104 Eq. 67

AF737295

DEC 04 1992

INTRODUCTION
TO
QUARTZ CRYSTAL UNIT DESIGN

Photograph of a bar of cultured quartz such as the kind produced commercially. (Courtesy of Sawyer Research Products, Inc.)

INTRODUCTION TO QUARTZ CRYSTAL UNIT DESIGN

by

Virgil E. Bottom, A.B., M.S., Ph.D.

Professor of Physics, Emeritus
McMurry College
Abilene, Texas

Van Nostrand Reinhold Electrical/Computer Science and Engineering Series

VNR VAN NOSTRAND REINHOLD COMPANY
NEW YORK CINCINNATI TORONTO LONDON MELBOURNE

Van Nostrand Reinhold Company Regional Offices:
New York Cincinnati

Van Nostrand Reinhold Company International Offices:
London Toronto Melbourne

Copyright © 1982 by Van Nostrand Reinhold Company

Library of Congress Catalog Card Number: 81-10494
ISBN: 0-442-26201-9

All rights reserved. No part of this work covered by the copyright hereon may be reproduced or used in any form or by any means – graphic, electronic, or mechanical, including photocopying, recording, taping, or information storage and retrieval systems – without permission of the publisher.

Manufactured in the United States of America

Published by Van Nostrand Reinhold Company
135 West 50th Street, New York, N.Y. 10020

Published simultaneously in Canada by Van Nostrand Reinhold Ltd.

15 14 13 12 11 10 9 8 7 6 5 4 3 2 1

Library of Congress Cataloging in Publication Data

Bottom, Virgil E.
Introduction to quartz crystal unit design.

(Van Nostrand Reinhold electrical/computer science and engineering series)
Bibliography: p.
Includes index.
1. Piezoelectric devices. 2. Oscillators, Crystal.
3. Quartz crystals – Electric properties. I. Title.
TK7872.P54B67 621.3815′2 81.10494
ISBN 0-442-26201-9 AACR2

To my teachers and to my students,
to all of whom I am indebted.

Van Nostrand Reinhold
Electrical/Computer Science and Engineering Series
Sanjit Mitra, Series Editor

HANDBOOK OF ELECTRONIC DESIGN AND ANALYSIS PROCEDURES USING PROGRAMMABLE CALCULATORS, by Bruce K. Murdock

COMPILER DESIGN AND CONSTRUCTION, by Arthur B. Pyster

SINUSOIDAL ANALYSIS AND MODELING OF WEAKLY NONLINEAR CIRCUITS, by Donald D. Weiner and John F. Spina

APPLIED MULTIDIMENSIONAL SYSTEMS THEORY, by N.K. Bose

MICROWAVE SEMICONDUCTOR ENGINEERING, by Joseph F. White

INTRODUCTION TO QUARTZ CRYSTAL UNIT DESIGN, by Virgil E. Bottom

FOREWORD

Quartz crystal units have been an item of commerce for more than a half century. The applications for quartz crystals have grown from their early roots in frequency standards and in amateur radio communication to the present wide spectrum of applications encompassing many electronic products. The millions of units produced annually are used extensively in the field of telecommunications and radio communications where precise frequency control, both in transmission and reception, is essential for quality and reliability of communication. Timekeeping was one of the early applications of quartz crystal units and has continued to grow in precision and in utilization. The earliest quartz clocks evolved into national standards of lapsed time and became the interpolation clocks for astronomical observations of epoch. As more fundamental atomic clocks evolved using a resonance line of cesium, quartz units continued to play a major role in providing the fly wheel for these quantum electronic clocks. In this application, the short term frequency stability of precision quartz resonators, which at present has achieved the level of a few parts in 10^{14} for one second sample times, provides again an excellent interpolation system for short times while the cesium resonance provides the absolute definition of frequency and time for periods longer than a few seconds. With the evolution of very small, rugged, low frequency, low cost resonators with improved temperature coefficients and long term stabilities approaching a part in a million per year, quartz crystal units have become practical as the frequency controlling element in electronic wrist watches. As this occurred, quartz watches rapidly replaced mechanical escapement watches and became the largest user of quartz crystal units. Their utility in timekeeping spans the entire spectrum from the lowest cost consumer product to the most precise time standard.

In addition to the applications of quartz crystal units in timing and communication, they have found extensive applications in other industrial products. They are used as timing control elements in computers, as surface wave correlators in digital communication, as discriminators and filters in instrumentation and communication, and as sensors of temperature and pressure.

Quartz crystal units provide frequency control over the range from a few kilohertz to a gigahertz. This frequency range of nearly a million to one is achieved by a variety of resonator designs from thickness flexure resonators through width flexure, extensional, face shear, thickness shear, and surface wave resonators. In all of these designs, the period of acoustic resonance is determined by the time of propagation of the acoustic wave through a resonant structure. At the lowest frequencies, the velocity of flexure waves is slowed through the use of very small dimensional ratios and the resonant path is extended to several centimeters. At the other extreme, the acoustic resonance dimensions are decreased into the submicrometer region where lithographic technologies set the upper frequency limit for surface wave devices.

Over the half century of quartz crystal unit development, there has been a steady growth in the frequency stability of resonators. The first international comparisons of frequency by W. G. Cady in the early 1930's indicated accuracies in the order of one tenth of one percent. Frequency stability of a part per million during World War II, and a part in 10^8 by 1952 were bench marks in the progress of precision. In the following decades, short term frequency stability improved to a few parts in 10^{10} by 1960, several parts in 10^{12} by 1970 and to a few parts in 10^{14} in 1980. This phenominal improvement in stability of approximately 10 orders of magnitude, over a period of 50 years, has evolved with relentless regularity at about an order of magnitude every 5 to 7 years.

Professor Bottom provides in this concise text the concepts and principles fundamental to the developments that have made possible extensive application of quartz crystal units over a wide range of frequencies with steady progress in performance. The latter chapters, 12 and 13, deal with the significant performance improvements that have evolved in quartz resonators. These chapters relate the concepts developed under the earlier discussions of crystallography, elasticity, the piezoelectric effect, and mechanical vibrations, to the device design and processing improvements that, in turn, have led to one of the most remarkable performance improvements of any device in the field of electronics.

Professor Bottom is well qualified to write this important text. He has conducted extensive academic research on piezoelectric resonators and lectured in the discipline underlying the technology of quartz crystal units. His experience in the Signal Corps Research Laboratories, as a director of industrial research and as a consultant to quartz crystal manufacturers, provides a complete perspective relating the theoretical concepts and analysis to the practical problems of device and process design.

Professor Bottom's book is a most welcome addition to the literature in the field of frequency and time.

Dr. Donald L. Hammond, Director
Physical Research Center
Hewlett Packard
Palo Alto, California

PREFACE

The quartz crystal unit is an indispensable component of modern electronic technology. Not only is it used to control and manage the frequencies in virtually all communication systems but it provides the isochronous element in most clocks, watches, computers, and microprocessors. It is a product of the discovery of the phenomenon of piezoelectricity by the Curie brothers in France just one century ago. Despite the importance, indeed the indispensability, of the piezoelectric resonator to the field of electronics only a few books have been written on the subject and none has heretofore been written with the specific objective of introducing the newcomer to the field of quartz crystal unit design. Nor has the subject of piezoelectricity found a place in the curriculum of schools of electrical engineering. Consequently few Electrical Engineering graduates have heard the word *piezoelectricity* and almost none has any familiarity with the basic principles involved. The result is a severe and growing shortage of people who are in a position to contribute to this important area of technology.

This book is presented with the hope that it will encourage some young engineers to become active in the field of applied piezoelectricity and help them to find their way into it.

Piezoelectricity is not a simple subject. The phenomenon is complex, involving the more advanced concepts of both electricity and mechanics. The applications of piezoelectricity involve almost every area of classical physics including mechanics, acoustics, wave motion, optics, electrostatics, electric circuit theory, x-rays and x-ray diffraction, and crystallography. Probably few fields of engineering or applied science require familiarity with a wider range of topics from the field of physics. The mathematical tools required include complex number theory, vector and matrix algebra, and some knowledge of partial differential equations. Fabrication and testing of quartz crystal units involve some of the most delicate and precise measurements which can be made. The product is an inexpensive component whose precision, reliability, and importance are not exceeded by any product of technology.

The range of physical and mathematical concepts required is wider than that covered in the usual undergraduate curriculum in either physics or Electrical Engineering departments, so beginning students often have difficulty with the works of Cady and Mason. It is hoped that this book will help bridge the gap between the usual undergraduate curriculum and these important works.

This book is avowedly tutorial. Those doing research in the field should turn to other sources. The purpose of this book is to help and encourage the beginning engineer to understand the basic principles of piezoelectricity and its applications to the design of quartz crystal units. For this reason, the mathematical derivations are simplified as much as possible. Complications such as second-order effects are deliberately omitted in order not to obscure the main topic (and not because they are unimportant). One-dimensional systems are used to illustrate basic principles. Anisotropy is sometimes ignored in order to simplify the discussion. The notation is simplified as much as possible in order not to obscure the main idea. The treatment is limited in scope; many important topics are not treated at all. Among the topics omitted are monolithic filter crystals, tuning fork and other low-frequency resonators, surface-wave devices, and most second-order effects. Almost all the book deals with thickness-shear-type resonators such as the AT-cut, but the principles developed can, once understood,be extended to much wider applications.

Most of the book is suitable for study by students who have completed good intermediate courses in general physics and differential and integral calculus. Where deemed necessary, some introduction to the mathematical tools is included. Derivations are given in a bit more detail than would be done in a more advanced text. The object is to elucidate rather than to abbreviate. A set of problems and questions is included as Appendix VI to help the student check and extend his understanding of the text.

This book is the result of many years of teaching the basic principles of piezoelectricity and quartz crystal unit design to undergraduate and graduate students in formal classes and to practicing engineers in seminars held under industrial auspices. Indeed the book began as a set of notes for such a seminar. It is now presented with the hope that it may serve a wider clientele.

Abilene, Texas | Virgil E. Bottom

SYMBOLS

The writer of a work of this type is faced with a dilemma in the choice of notation. One approach is to use the symbols which have conventionally been employed to represent the numerous quantities. The other is to set up a completely consistent notation with no ambiguities. The first approach results in the use of the same symbol for different quantities. The second approach leads to proliferation of symbols and difficulty in referring to earlier works. The author has chosen the first approach, believing that in instances where a given symbol represents different quantities, the meaning will be clear to the reader from the context.

A	amplitude, area
a	distance, radius, rotation matrix
b	velocity ratio, rotation matrix
C	capacitance, motional capacitance (also C_m)
C_o	static capacitance
c	stiffness coefficient, rotation matrix
D	electric displacement
$D\theta, D\phi$	angle between atomic plane and crystal plate
d	piezoelectric strain coefficient, atomic spacing, plate diameter
E	electric-field intensity
e	piezoelectric stress coefficient, thickness, voltage, extension
F	force
f	frequency
g	goniometer reading, acceleration of gravity
h	spring stiffness
I	moment of inertia, electric current
i	current
j	$(-1)^{1/2}$
K	frequency constant
k	dielectric constant, coupling coefficient, mode index, $4\pi\epsilon_o$, $2\pi/\lambda$
L	length, inductance, motional inductance (also L_m)
l	length, mode index

m	meter, mode index
N	Avagadro's number
n	mode index, harmonic order
P	polarization
p	dipole moment
Q	quality factor
q	electric charge, q_v space charge density
R	Resistance, motional resistance (also R_m), radius of curvature
r	damping coefficient, distance, capacitance ratio
S	frequency-stress coefficient
s	compliance coefficient
T	period, time, temperature
T_f	frequency-temperature coefficient
t	time
U	internal energy
V	electric potential
v	wave velocity
w	width
X	stress, reactance
x	strain, coordinate variable
Y	admittance
y	coordinate variable
Z	impedance
α	rotation matrix, coefficient of linear expansion
β	imaginary part of propagation constant, angle of deflection of x-ray beam
γ	complex propagation constant
δ	angle between atomic planes and surface of blank
	partial derivative
Δ	incremental change, $r/2Q^2$
ϵ	piezoelectric stress coefficient (also e)
ϵ_o	susceptibility of space
η	dielectric susceptibility
κ	coupling coefficient
λ	wavelength
ρ	density
σ	surface charge per unit area
θ	angle
θ_B	Bragg angle
ϕ	angle
χ	root of Bessel's function of the first kind
ψ	mechanical displacement, azimuth angle
ω	angular frequency

CONTENTS

INTRODUCTION TO QUARTZ CRYSTAL UNIT DESIGN

1
INTRODUCTION

THE ELECTROMAGNETIC SPECTRUM

The radio frequency spectrum extends from about 15,000 cycles per second (15 kHz) to at least 100,000,000,000 cycles per second (100,000 MHz). Over this entire range of frequencies, it is essential that the frequency be accurately controlled to avoid interference between transmitters operating in the same geographical area or on adjacent channels. The continuing demand for space in the available radio-frequency spectrum has resulted in closer and closer spacing of the assigned frequencies with accompanying requirements for better frequency control. Maximum utilization of the limited frequency spectrum requires the highest possible stability of the transmitted frequency and the maximum selectivity at the receiver.

The use of single-sideband and suppressed-carrier communication systems allows more communication channels in a given frequency band but imposes severe requirements on the frequency stability of both transmitter and receiver. Many navigation and ranging systems depend upon precise frequency control and accurate time measurements.

Over much of the frequency spectrum, all these purposes are best accomplished by the use of quartz crystal units as active elements in oscillator circuits and as passive elements in electrical wave filters.

OTHER APPLICATIONS OF THE QUARTZ CRYSTAL UNIT

The ability of the vibrating quartz element to mark off equal intervals of time makes it a natural choice as the isochronous element in timekeeping devices. The development of suitable electronic components to use with the vibrating quartz element now makes quartz clocks and watches practical and it appears that the torsional pendulum which has been the basis of the watch industry for over three centuries has now been superceded by the quartz piezoid.

The temperature-frequency dependence of a specially designed quartz piezoid is used in a thermometer of great convenience and accuracy. The quartz thermometer is capable of providing almost instantaneous digital temperature information from remote locations. The frequency-stress dependence of a vibrating piezoid of quartz has been used as a pressure indicator and as an accelerometer. New applications of the vibrating quartz piezoid are continually being found and doubtless many remain yet undiscovered.

THE PIEZOELECTRIC EFFECT

The word *piezoelectricity* literally means "pressure electricity"; the prefix *piezo* (pronounced pi-e-zo) is derived from the Greek word *piezein*, "to press." The word has been generally defined by Cady as follows: "piezoelectricity is electric polarization produced by mechanical strain in crystals belonging to certain classes, the polarization being proportional to the strain and changing direction with it." The production of an electric polarization by mechanically inducing a strain in a crystal is called the direct piezoelectric effect. The converse effect, whereby a mechanical strain is produced in a crystal by a polarizing electric field, also exists. The converse piezoelectric effect has sometimes been confused with the electrostrictive effect which occurs in solid dielectrics such as glass. The two effects differ in two important respects. The piezoelectric strain is usually larger by several orders of magnitude than the electrostrictive strain, and the piezoelectric strain is proportional to the electric-field intensity and changes sign with it, whereas the electrostrictive strain is proportional to the square of the field intensity and therefore independent of its direction. The electrostrictive effect occurs simultaneously with the piezoelectric effect but (at least in quartz) may be ignored for practical purposes.

In some crystals, electric polarization is produced by simply squeezing the crystal along a certain axis, for example, by clamping it in a vice. Other types of strain such as bending, shearing, and torsion also produce polarization in crystals of certain types. Often several effects occur simultaneously. Conversely, the application of polarizing electric fields may cause the crystal to experience a longitudinal or shearing stress and under certain conditions bending, torsion, and flexure may be produced. This is called the *converse piezoelectric effect.*

The term piezoelectricity is also used with reference to materials such as barium titanate in which electric fields produce mechanical strains. These materials are polycrystalline, however, and have a permanent dipole moment induced in them during manufacture. They also exhibit a domain structure. Their behavior is analogous to that of ferromagnetic materials in which magnetic fields produce mechanical contraction called *magnetostriction.* These materials are not considered in this book.

CRYSTALS

In the physical sense, a crystal is a set of atoms, molecules, or ions arranged in a definite geometric pattern in three dimensions. The ideal crystal is made up of unit cells which are the smallest parallelepipeds that can be fitted together to form the crystal. Each unit cell of a given crystal is identical to every other unit cell of the crystal in that each one contains the same set of atoms, molecules, or ions, arranged in exactly the same way.

The atoms, molecules, or ions which make up the crystal are arranged in rows, columns, and planes. The natural faces and the cleavage planes of the crystal tend to be parallel to planes of atoms. The distances between the planes of atoms and the dihedral angles between the nonparallel planes are as characteristic of the crystal as its density or its chemical composition. Crystallographers long ago deduced much about the internal structure of crystals by studying their external forms.

Crystallographers recognize 32 classes of crystals of which 20 exhibit the piezoelectric effect. It is thus apparent that piezoelectricity is not a particularly rare phenomena. Although many crystals exhibit the piezoelectric effect, very few are useful, and quartz alone provides the necessary combination of mechanical, electrical, chemical, and thermal properties required for making piezoelectric elements for the electrical communication field.

QUARTZ

Crystalline quartz is composed of two elements, silicon and oxygen. The chemical formula is SiO_2. In the amorphous form SiO_2 is a major constituent of many rocks and most sand. The crystalline form of SiO_2, or quartz, is also relatively abundant in nature; but in the highly perfect form required for the manufacture of quartz crystal units, the supply is not large. Most of the high-quality crystals of natural quartz are obtained from Brazil. Some domestic sources of quartz are known, but the low quality and high cost of production make their use prohibitive.

Natural quartz crystals were grown from water solutions containing dissolved SiO_2. The crystals were probably grown at moderate temperatures over long periods of time. Sometimes the crystals are found in their original formations but more often the cavities, or "vugs," in which the crystals grew, have weathered away and the harder quartz crystals are now found in the debris or in river beds where they have been deposited.

The limited supply and the high cost of natural quartz have resulted in the development of a cultured quartz industry. By improving on the work done in Germany half a century ago, it is now possible to produce crystals of quartz which can be used in most applications.

Silicon dioxide is only slightly soluble in water at ordinary temperatures. However, in alkaline solutions at elevated temperatures, the solubility is increased appreciably. Crystals of cultured quartz are grown by dissolving SiO_2 in an alkaline solution of water at temperatures near the critical temperature of water. Very strong autoclaves are required to withstand the pressures, which may exceed 1000 atm. Seed crystals are placed in the cooler part of the autoclave and nutrient in the form of chunks of SiO_2 is placed in the warmer portion. The temperatures and temperature gradients are controlled by computer programs. The solution moves by convection from the hotter to the cooler region dissolving the nutrient and depositing on the seed crystals. Crystals having masses of several hundred grams can be grown in a few weeks. The quality of the quartz is critically dependent upon the conditions of growth; in general, the slower the growth, the higher the quality. Most of the quartz processed in the United States (1980) is cultured (man-made) quartz. The crystals are grown in such shapes and sizes that the labor and loss of material in processing are minimized.

The quality of the quartz for piezoelectric use can be predicted by measurements of its infrared absorption. It is customary to inspect the quartz in this way and to grade it for specific applications. Natural quartz is still preferred for a few of the most severe requirements. The differences between samples of quartz crystals are due to crystalline imperfections. These may include such defects as lattice vacancies, interstitial atoms, impurity atoms, and growth dislocations. The common defects of twinning in natural quartz are not present in cultured quartz.

The crystalline form of SiO_2 at temperatures below 573°C is called *alpha quartz*, or simply *quartz*, and it is with this material that we will be exclusively concerned. Much of the information may be applicable to other materials, of course, but the direct application will be to the use of quartz in the design and fabrication of oscillator and resonator elements. No distinction is required between natural and cultured quartz and none will be made in the material which follows.

HISTORICAL REVIEW

The discovery of the piezoelectric effect is usually ascribed to Jacques and Pierre Curie, who announced their discovery in 1880. Not only did they demonstrate the phenomenon, but they also established the criteria for its existence in a given crystal. The Curie brothers measured the values of the piezoelectric constants in several crystals, including quartz, and devised a few instruments utilizing the phenomena.

Except for an isolated instance here and there, the phenomenon of piezoelectricity was not exploited until the period of World War I, when Langevin (France) used piezoelectrically excited quartz plates to generate sound waves in water for use in submarine detection. Langevin later used quartz plates as receivers of sound waves, thus employing both the direct and converse piezoelectric effects. Langevin's work, of course, was the forerunner of sonar. Interest in piezoelectricity was revived by the work of Langevin, and among those who became interested in the field were Cady at Wesleyan University and Nicholson of the Bell Telephone Laboratories. Their work led to the development of the piezoelectric resonator as a frequency-controlling element.

An oscillator employing a Rochelle salt crystal was patented by Nicholson of the Bell Telephone Laboratories in 1918. It is not altogether clear whether or not he conceived the function of the crystal as being primarily a frequency-controlling element; but in any case, the control could not have been very effective. In 1920 Cady patented an oscillator circuit, the frequency of which was definitely controlled by a quartz resonator. Cady's oscillator consisted of a three-stage amplifier with the quartz resonator in the feedback path. A little later, Pierce showed that a crystal-controlled oscillator could be built using only one vacuum tube.

The first broadcast station to be crystal-controlled was WEAF in New York City. It was placed in operation in 1926, and soon thereafter most broadcast stations began to utilize quartz crystals for frequency control. Throughout the period from 1926 to 1939, quartz crystal units were used by amateurs, broadcast stations, and some manufacturers of two-way radio-communication equipment. These units were made by a small number of suppliers, often on a custom basis. No industry, as such, existed.

The use of crystal control in military communication equipment did not become common until immediately prior to World War II. The armed services were in the process of converting to crystal control when the United States entered the war. In 1940 it was estimated that in the event of a major war, the armed services might require as many as 100,000 units. This seemed like a fantastic quantity at a time when all crystal units were made in a few small shops where a skilled worker might be able to make as many as 10 units in a day. During the war, however, over 30 million quartz crystal units were produced through a crash program costing over $1 billion. At the peak of the war effort over 125 factories were engaged in the production of quartz crystal units for the armed services of the United States and the Allies.

THE PIEZOELECTRIC RESONATOR

The first quartz resonators were simply quartz plates cut from the crystal in such a way that the normal to the plate is parallel to the *X*-axis of the crystal.[1] Such a piezoid, called an *X*-cut, has a temperature coefficient of frequency of about –20 Hz/(MHz)(°C). In other words, the frequency decreases 20 Hz/MHz for each degree rise in temperature.

The *Y*-cut plate, which was introduced by E. D. Tillyer in the late 1920s, has certain advantages over the *X*-cut plate. It is only about two-thirds as thick as an *X*-cut plate of the same frequency. It also has fewer undesirable modes of vibration and is less affected by air damping. Fabrication is also somewhat simpler. Its most important advantage is that it can be clamped for mounting. The *Y*-cut has two disadvantages, however. The temperature coefficient of frequency is about +100 Hz/(MHz)(°C) and the device often refuses to operate or changes frequency abruptly as the temperature is changed.

It was discovered by groups in Germany, Japan, and America in 1929 that the temperature coefficient of the *Y*-cut plate could be improved and in fact made to become zero at certain temperatures by rotating the plane of the *Y*-cut about the *X*-axis. The first of the zero temperature coefficient cuts to be discovered was named the *AT-cut.* It is still the most widely used of all the quartz piezoids. The second zero-temperature-coefficient orientation to be discovered was named the *BT-cut.* The BT-cut, which is 50 percent thicker than the AT-cut for a given frequency, was widely used during World War II, but it has been largely replaced by the AT-cut, which can be made to have a smaller frequency excursion over the wide temperature ranges required by modern communication systems.

Quartz piezoids can be made in many different forms. Quartz bars, vibrating in flexure modes (like xylophone bars), are used from the audio range to some tens of kilohertz. Such elements are used as the isochronous elements in wristwatches. Small piezoids in the form of tuning forks, operating around 30 kHz, have also been developed for use in wristwatches. Quartz bars vibrating longitudinally

---

[1] The terms *X-cut*, *X-axis*, etc., are defined in Chap. 2.

and quartz rings vibrating radially have been used as frequency standards at frequencies from 50 to 100 kHz. In the range from 100 to 1000 kHz, quartz plates vibrating in the face-shear mode have been widely used.

Piezoids of many other configurations have been developed for special applications. Most are designed to make the frequency as nearly independent of temperature as possible. Most of the material covered in this book relates directly to piezoids of the thickness-shear mode type, i.e., Y-cut, AT-cut, and BT-cut plates. However, with suitable interpretation nearly all the material is applicable to piezoids of any orientation.

FABRICATION TECHNIQUES

In the early days, quartz crystal units were fabricated by hand methods which were very crude and primitive in comparison with present-day manufacturing techniques. Originally the quartz plates were cut with "muck saws," which were simply metal discs running in a slurry of abrasive and oil or water. The orientation of the plates was determined by reference to the natural faces of the quartz crystal. Orientation was therefore inaccurate, and rarely were two finished units found to perform in the same way. The quartz plates were mounted between two metal plates which served as electrodes and also to support the quartz plate.

During the past four decades, enormous improvements have been made in fabrication techniques. The muck saw was replaced by the diamond saw, which is capable of cutting thin wafers of quartz with extraordinary speed and accuracy. The use of x-ray diffraction replaced orientation by reference to natural faces. Precision machinery for lapping and controlling the dimensions of the plates with precise orientation now make it possible to fabricate units to a given design with such accuracy that the mechanical and electrical characteristics are substantially identical.

In the past few years a new type of slurry saw has replaced the diamond saw, especially for sawing cultured quartz. This saw consists of a set of 100 or more fine, stretched metal bands moving back and forth across a number of bars of quartz in a flood of abrasive slurry. These "gang saws" are capable of cutting several hundred crystal blanks per hour with very little cost for labor.

The electrodes of modern crystal units are metal films deposited directly on the quartz by vacuum evaporation. Not only does this provide a more convenient method of adjusting the frequency but it also eliminates the air gap between the quartz and the electrode which is a principle source of difference between units of a given type. A further advantage is that the units are more rugged and less susceptible to the effects of shock and vibration. The units are also lighter and less expensive.

While improvements in the temperature coefficients, mechanical stability, and reliability of the quartz crystal unit were being made, the frequencies realizable were also being increased. Prior to World War II most crystal units operated in or a little above the broadcast band. A few units operating at frequencies of 3 to 4 MHz were made for "short-wave" enthusiasts. During the war the practical limit was pushed to approximately 10 MHz but not without serious difficulties, especially with "aging." Today (1980) it is possible to purchase reliable crystal units operating at frequencies above 150 MHz. This appears to be approaching the upper practical limit for direct control of frequency by the use of vibrating plates. Direct methods of controlling frequencies in the gigahertz range with the precision of quartz resonators are sorely needed. Recent developments in the area of surface (Rayleigh) waves may result in devices which will help to meet this need.

SUMMARY

The quartz crystal unit is one of the most remarkable devices made by human beings. Whether considered from the standpoint of the complexity of the physical phenomena involved; the delicate operations required in the fabrication of the device; the unique properties of the basic material, the precision, reliability, or low cost of the device; the quartz crystal unit holds a place which is almost unique among manufactured articles.

A typical crystal unit is specified to operate at a frequency of 50 MHz over a temperature range of −55 to +105°C with a maximum frequency excursion of ±0.005 percent of its nominal frequency. It is specified to operate continuously under severe conditions of shock or vibration in any position over long periods of time. It may not be generally appreciated that a tolerance of ±0.005 percent is equivalent to about 4 sec in 24 hr or about 3 in/mi (or 5 cm/km). Yet the typical crystal unit is expected to maintain this precision under any combination of the adverse conditions described while executing 50 million mechanical vibrations per second. Furthermore, it is required to continue this performance over long periods of time without change of frequency or failure. The number of mechanical vibrations executed by a 12-MHz quartz plate reaches the astounding total of 10^{12} per day!

Note the forces involved in the vibration of the quartz. The actual displacement of a point on the surface of the quartz is, of course, quite small – generally only a few atomic distances. Even this small displacement results in tremendous forces, however. It can easily be shown[2] that a point on the surface of a quartz plate vibrating at 50 MHz experiences accelerations of several million "g's." It can also be shown[3] that the loss of an amount of quartz from the surface of the plate, equivalent in thickness to one layer of atoms, would result in an increase of several hundred cycles per second in the resonant frequency. This is an appreciable part of the total frequency change allowed for all effects including temperature and aging.

It is indeed fortunate for the communication art that nature has provided a material with the piezoelectric, mechanical, and electrical properties required to meet such exacting conditions.

It is not surprising that in order to achieve such results in a manufactured article, a high order of scientific and engineering skill plus meticulous care at every step in the fabrication process are required.

Considering all the factors involved, it appears that quartz, in the words of W. P. Mason, will remain "the cornerstone of the communication art."

---

[2] The displacement of a point vibrating with simple harmonic motion may be described as $x = A \sin(\omega t)$, where A is the amplitude and ω is the angular frequency. The acceleration $\ddot{x} = -\omega^2 A \sin(\omega t)$. Hence, the maximum value of the acceleration is $\omega^2 A$. If we assume that A is only one lattice spacing, i.e., about 5×10^{-10} m and $\omega = 2\pi f = 3 \times 10^8$ sec^{-1}, then the acceleration is 45×10^6 m sec^{-2}, or about 5 million 'g's.

[3] The resonant frequency of a plate vibrating in a thickness-shear mode is given by $f = nv/2e$, where e is the thickness, v is the wave velocity, and $n = 1, 3, 5, \ldots$. Differentiating the expression for f gives $df = -(f/e)de$. A 50-MHz plate vibrating in its third-overtone mode ($n = 3$) has a thickness of approximately 1.0×10^{-2} cm. A change of thickness of 5×10^{-8} cm (about one atomic layer) therefore results in a change of frequency of about 250 Hz.

2 THE PHYSICAL CRYSTALLOGRAPHY OF QUARTZ

In order to deal with the problems of designing, developing, and fabricating quartz crystal units, it is necessary to have some familiarity with crystal physics. Only those portions of this very extensive field which are required to design quartz piezoids or to understand their operation will be discussed. For complete treatments of the subject, the reader may consult one or more of the works listed in the bibliography at the end of the book.

ANISOTROPY

Crystals differ from noncrystalline solids (if such materials may be called solids) in that many of the physical properties of the crystal are dependent upon direction in the crystal whereas in a noncrystalline or amorphous material, the properties are independent of direction. For instance, one may speak of the hardness of glass without qualification; but in speaking of the hardness of quartz, the direction must be specified, since this property, like many others, differs in different directions in the crystal. If the coefficients describing the physical properties of a solid differ with direction, the solid is said to be anisotropic. A striking example of anisotropy is the electrical conductivity of quartz, which is about 1000 times greater in one direction than another in the crystal. Except for a few properties such as density, specific heat, melting point, etc., most properties of crystalline materials are direction-dependent. Table 2.1

Table 2.1 Some Anisotropic Properties of Crystals

Elastic modulus and compliance
Thermal conductivity
Electrical conductivity
Coefficient of thermal expansion
Hardness
Solubility
Dielectric constants
Piezoelectric constants
Index of refraction (optical)
Velocity of propagation of longitudinal (sound) waves
Velocity of propagation of shear waves

shows a few of the physical properties of crystals, the values of which, in general, depend upon direction in the crystal.

THE QUARTZ CRYSTAL

Since we are concerned exclusively with quartz in this book, we should first become familiar with the general form of the quartz crystal. The idealized quartz crystal is a hexagonal prism with six cap faces at each end, as shown in Fig. 2.1. The prism faces are designated m-faces and the cap faces are designated R- and r-faces. The R-faces are often called major rhomb faces and the r-faces are called minor rhomb faces. There are, of course, three of each at each end of the complete crystal. These faces betray the trigonal structure of quartz and its threefold symmetry.

Quartz is an enantiomorphic crystal, i.e., it exists in right and left form. The faces designated s and x are rarely seen on crystals, but when present, they enable one to distinguish between right and left quartz. Both kinds of quartz are equally useful for fabricating piezoids, but a distinction is required in processing the material.

Crystals of cultured quartz exhibit the same faces as do crystals of natural quartz, but they are sometimes a little difficult to identify. We will return to this topic after discussing the subject of crystal axes.

CRYSTAL AXES AND AXIAL SYSTEMS

Since the values of the coefficients describing the physical properties of the crystal are in general dependent upon direction, it is necessary

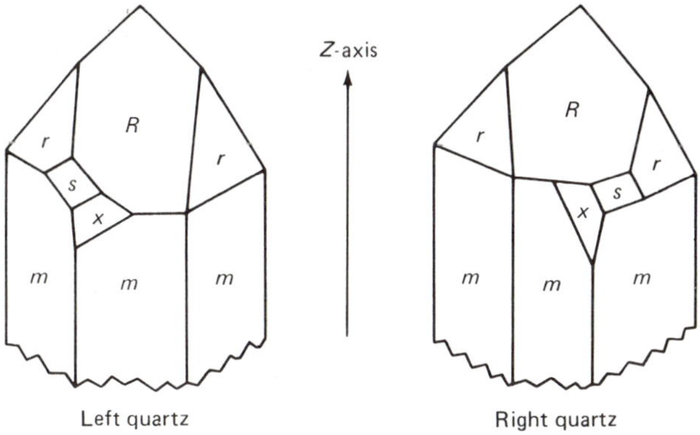

Fig. 2.1. The quartz crystal. Crystals with cap faces on both ends are rare. The *s* and *x* faces are also rare. In right quartz, the *s* and *x* faces occur at the lower-right corner of an *R* face. In left quartz, they occur at the lower-left corner.

to choose reference directions within the crystal for use in specifying their values. Such directions are called the *crystal axes.* It must be very carefully noted that a crystal axis is not a line but a direction in the crystal. The axial directions are chosen to make the description of the crystalline properties as simple as possible. An axis is always chosen with reference to some specific physical property of the crystal, which in turn depends upon the way in which the atoms are arranged within the crystal.

Different axial systems may be used to describe the same crystal, so confusion may result unless it is made quite clear which system is being employed. Furthermore, one axial system may be most convenient for one purpose and another system for another purpose. For example, it is more convenient to use the Bravais-Miller axial system to specify the natural faces and atomic planes in quartz, whereas it is more convenient to use an orthogonal coordinate system for purposes of computations involving the piezoelectric and mechanical properties of the crystal.

The Bravais-Miller (B-M) system of axes consists of a Z-axis (sometimes called the c-axis) and three X-axes (sometimes called a_1, a_2, a_3) which make angles of 120° with each other and which lie in a plane perpendicular to the Z-axis. The B-M system illustrated in Fig. 2.2*a* is applicable to crystals such as quartz which have trigonal symmetry. The use of the B-M system in specifying atomic planes is discussed in the next section.

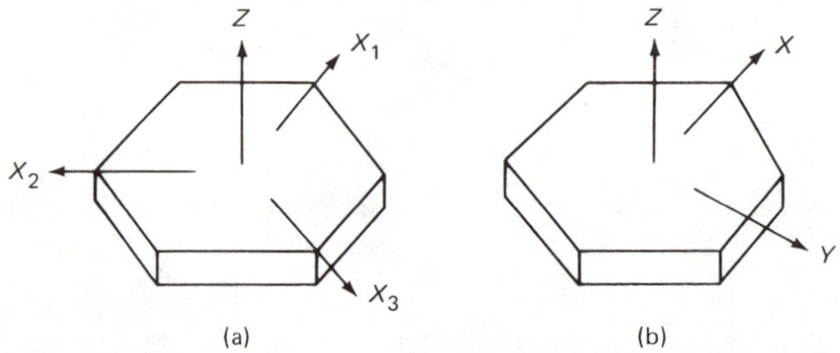

Fig. 2.2. **Cross sections of a quartz crystal taken perpendicular to the *Z*-axis, showing (*a*) the B-M system of axes and (*b*) the rectangular system.**

The rectangular, or orthogonal, coordinate system of axes is shown in Fig. 2.2*b*. The *Z*-axis in the rectangular system is the same as the *Z*-axis in the B-M system. The *X*-axis of the rectangular system is one of the three *X*-axis of the B-M system. It is immaterial which of the three is chosen since they are identical. The *Y*-axis is perpendicular to both the *Z*- and the *X*-axes. The method of specifying the sense of the *X*-axis will be discussed later.

THE Z-AXIS AND OPTICAL TWINNING

The principle, or *Z*-axis (also called the optic axis and the *c*-axis) in quartz is an axis of threefold symmetry. This means that all the physical properties repeat each 120° as the crystal is rotated about the *Z*-axis. If a beam of plane-polarized light is passed through the crystal in the direction of the *Z*-axis, the plane of polarization is rotated by 15 to 45° for each millimeter of quartz traversed. The exact amount of rotation depends on the color of the light and the temperature of the quartz. This phenomenon is called optical rotation and it forms the basis of a number of optical instruments. It has no special significance in the field of piezoelectricity.

In some specimens of quartz the plane of polarization is rotated one way and in other specimens the rotation is in the opposite sense. The crystal is said to be right-hand (or simply right) quartz if the plane of polarization is rotated clockwise as seen by an observer looking through the quartz toward the source of light. Left quartz produces the opposite rotation. Both types of quartz may be used in fabricating piezoids, but some attention must be given to the hand in cutting certain types of piezoids.

Most crystals of natural quartz are composed partially of left and partially of right quartz. Such crystals are said to be *optically twinned*. The presence of optical twinning can be detected by inspection with polarized light or by etching a surface with hydrofluoric acid or ammonium bifluoride. Due to anisotropy the rate at which the acid attacks the quartz is different in different directions. Consequently, small etch pits having the form of a shingled roof begin to develop. The etch pits have different orientations in the two parts of the crystal and therefore reflect light differently, revealing the boundary between the right and left portions. Figure 2.3 illustrates the appearance of a Z-section of twinned quartz after etching.

Piezoids can be fabricated from either right or left quartz, but it is usually not permissible to have both types present in the same one. For this reason it is necessary to inspect the quartz carefully to ensure that no twin boundaries are present. Cultured quartz is usually free of optical twinning. The seed crystals are carefully selected to be free of twinning and other defects, since any defect present is usually propagated through the crystal as it grows.

THE X-AXIS AND ELECTRICAL TWINNING

The X-axes in quartz are parallel to a line bisecting the angles between adjacent prism faces. The X-axis is also called the *electrical axis*. More precisely it is a polar axis, since electric polarization

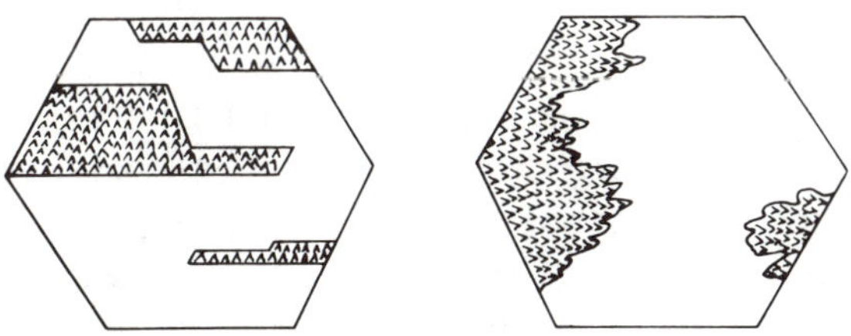

(a) (b)

Fig. 2.3. ***(a)*** **Optical twinning and** ***(b)*** **electric twinning. The boundaries between optical twins are usually straight lines. The boundary between electrical twins are usually irregular.**

occurs in this direction upon mechanical strain. Only those crystals having one or more polar axes display the phenomenon of piezoelectricity.

If a plate of quartz is cut in such a way that the major surfaces are perpendicular to the X-axis, the resulting piezoid is called an X-cut. If the plate is subjected to mechanical compression, a positive charge is developed on one side and an equal negative charge on the other. These charges are readily observed by placing conducting films on each side of the plate and connecting these to a direct-current vacuum-tube voltmeter. If the input capacitance of the meter is low, potential differences of several volts result from the application of a pressure of a few newtons per square centimeter.

The sign of the charge developed alternates between positive and negative at successive corners of the crystal. Following the IEEE standards, the positive X-direction is chosen to be the direction in which a positive strain produces a positive charge. A positive strain is defined to be the extension resulting from tension. By this definition, that end of the X-axis is positive where a negative charge is developed by compression.

Quartz crystals are often electrically twinned. Electric twinning consists of a reversal of the X-axis. In an electrically twinned crystal compression produces a positive charge on one side of the twin boundary and a negative charge on the other. Electric twinning (also known as *Dauphiné twinning*) usually occurs in irregular patches within the crystal; and the boundaries, revealed by etching, do not conform to any geometric pattern, as illustrated in Fig. 2.3*b*.

At temperatures above 573°C, called the *Curie*, or *inversion*, temperature, quartz exists in a hexagonal form called *beta quartz*. The sense of the polar axis is lost when the crystal changes from the alpha to the beta phase. When the crystal is cooled through the inversion temperature, it returns to the alpha phase, but the sense of the polar axis is usually opposed in different parts of the crystal, i.e., the crystal is electrically twinned. Electrical twinning can be induced in quartz at temperatures below the inversion temperature by mechanical stress. It is therefore important that thermal and mechanical stresses be avoided in processing quartz for piezoelectrical applications.

Cultured quartz is usually free of electrical twinning, but it may occur if the seed crystal is twinned or if undue stress is exerted on the crystal during or after growth. A twinned crystal can often be identified by a "shingled" appearance on the $+X$-surface.

THE Y-AXIS

The Y-axis is not a polar axis in quartz and no electric polarization results from an extensional strain in the Y-direction. It will be seen later, however, that a shear stress applied to a Y-cut quartz plate does result in electric polarization. This relationship forms the basis of a number of important types of quartz piezoids.

SPECIFICATION OF ATOMIC PLANES IN CRYSTALS

In order to describe the position of a given set of parallel atomic planes in a crystal, a coordinate system is required. In crystals of cubic symmetry such as the alkali halides (NaCl, KCl, etc.), the ordinary XYZ coordinate system is used. The directions of the coordinate axes are chosen to be parallel to the natural cleavage planes. Any plane in the crystal must intercept at least one of the coordinate axes. In case the plane is parallel to the axis, it is said to intersect it at the point of infinity. The Miller indices of the plane is a set of three numbers which are the smallest integers proportional to the reciprocals of the intercepts of the plane with the X-, Y-, and Z-axes, respectively.

Suppose, for example, we imagine a plane to intercept the X-axis at a distance $2a$ from the origin. Let it intercept the Y-axis at $3a$ and the Z-axis at $1a$. The reciprocals of these three numbers are, respectively, $1/2a$, $1/3a$, $1/1a$. Multiplying each of the numbers by $6a$, we arrive at the set of numbers 3, 2, 6. These numbers, written (326) and read "three, two, six," are the Miller indices of the plane.

As another example, consider a plane which intersects the X-axis at $1a$, the Y-axis at $3a$ and is parallel to the Z-axis. The intercepts are $1a$, $3a$, and ∞, respectively. The reciprocals are $1/1a$, $1/3a$, 0. Multiplying each of these numbers by $3a$ gives the Miller indices (310). The plane is called the *three, one, zero plane*. Any plane parallel to this plane is also a (310) plane, so that the Miller indices of a plane do not refer to any particular plane of atoms but to all members of a set of parallel atomic planes.

In order to specify planes in a crystal such as quartz, it is necessary to use the Bravais-Miller indices and the trigonal axial system. The *Z*- axis and the three *X*- axes form a coordinate system. Any plane in the crystal intersects each of the four coordinate axes including the point at infinity when the plane is parallel to one of the axes. The Bravais-Miller indices then constitute a set of four numbers which are the smallest integers that are proportional to the reciprocals of the intercepts of the plane with the $X_{\bar{1}}$, $X_{\bar{2}}$, $X_{\bar{3}}$, and *Z*-axes respectively.

As an illustration of the use of the B-M indices to describe a crystal face or plane, consider the *m*, or prism, face of a crystal of quartz. Figure 2.4 is intended to show a *Z*-section i.e., a surface normal to the *Z*-axis. The *m* planes are therefore perpendicular to the plane of the page. If we call *a* the unit distance along the *X*-axes, the intercepts of plane 1 with the $X_{\bar{1}}$, $X_{\bar{2}}$, $X_{\bar{3}}$, and *Z*-axes are a, ∞, $-a$, and ∞, respectively. The B-M indices are therefore proportional to the numbers $1/a$, 0, $1/-a$, and 0 or 1, 0, −1, and 0. The B-M indices are therefore (1 0 −1 0). In the B-M system the algebraic sum of the first three indices must be zero, so that writing the third digit is superfluous and it is often replaced by a dot. The B-M indices of the plane are then (10.0). The plane we have just studied is marked 1 in Figure 2.4. The other *m*-faces, marked 2 and 3, have different indices which can be determined in the same way. When a

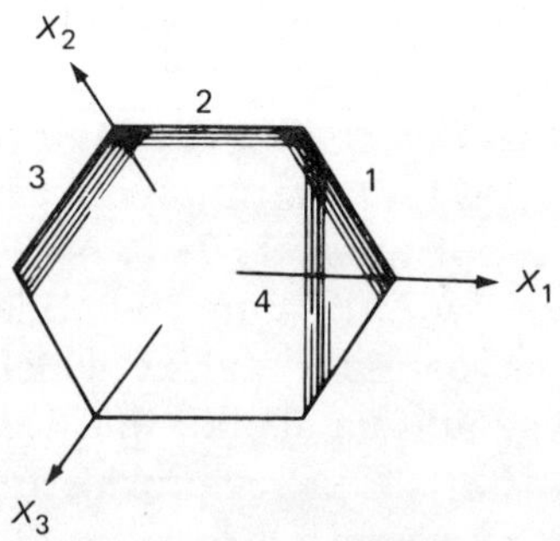

Fig. 2.4. A *Z*-section of a crystal of quartz. The *Z*-axis is perpendicular to the page.

minus sign appears in the B-M indices it is usually written by putting the minus sign above the numeral. Hence, the B-M indices for the planes marked 3 which are (–1 1 . 0) becomes $(\bar{1}1.0)$.

All the other atomic planes may be specified in the same way. For example, the set of planes normal to the *X*-axis which are designated as 4 in Fig. 2.4 intercept the axes at $a/2$, $-a$, $-a$, and ∞, respectively. Therefore, the B-M indices of these planes are (21.0). Although the *X*-planes do not form a natural face in quartz, they are important because a number of piezoids are prepared with their surfaces parallel to these planes. In fact, the first piezoelectric devices made by the Curies were *X*-cut plates of quartz.

The B-M indices for each of the three *m*, *R*, *r*, *s*, and *x* natural faces are given in Table 2.2.

The plane perpendicular to the optic or *Z*-axis is designated the (00.3) plane. It is not a natural face. The AT-cut is oriented with respect to one of the *r*-faces with which it makes a dihedral angle of about 3°. Another important set of planes which do not form a natural face are the (20.3) planes which make an angle of about 49° with the *Z*-axis and are used in orienting the BT-cut. Similarly, the (12.2) planes are used in orienting the IT- and SC-cuts.

Tables of many atomic planes which are useful as reference planes by which to orient quartz piezoids are given by Heising.[1] A few of the more commonly used planes are listed in Table 11.2. (p.183)

The enantiomorphic nature of quartz makes it necessary to consider the hand of the crystal in setting up a coordinate system. Fortunately the values of the coefficients of all the physical properties and the mathematical relationships between them become identical

Table 2.2 The B-M Indices of the Natural Faces in Quartz

	m	*R*	*r*	*s*	*x*
1	(10.0)	(01.1)	(10.1)	(11.1)	(51.1)
2	$(0\bar{1}.1)$	$(\bar{1}0.1)$	$(\bar{1}1.1)$	$(1\bar{2}.1)$	$(\bar{6}5.1)$
3	$(\bar{1}1.0)$	$(1\bar{1}.1)$	$(0\bar{1}.1)$	$(\bar{2}1.1)$	$(1\bar{6}.1)$

[1] R. A. Heising, *Quartz Crystals for Electrical Circuits.* New York: Van Nostrand, 1946, p. 105.

if a right-hand coordinate system is used for right crystals and a left-hand coordinate system with left crystals. For this reason the IRE (now IEEE) Standards Committee has recommended that the following conventions be used with quartz crystals.

1. A right-hand coordinate system is used to describe the physical properties of a crystal of right quartz and a left-hand system for left quartz.
2. In a right-hand coordinate system a positive rotation is one which appears counterclockwise when observed from the positive end of the axis of rotation.

Because of the peculiar screwlike nature of the structure of quartz, it is immaterial which end of the *Z*-axis is chosen as positive. This is analogous to the easily confirmed observation that a right-hand screw is not changed to a left-hand screw by simply looking at it from the opposite end.

CULTURED QUARTZ

Most of the quartz used in the quartz crystal industry today (1980) is man-made or cultured quartz. (The term *artificial quartz* is to be avoided, since all quartz is crystalline SiO_2, whether crystallized in nature or in a factory.) Although the cost of cultured quartz per kilogram is generally higher than that of natural quartz, the cost per crystal unit is usually less. And in any case, the supply of natural quartz is inadequate to meet the demands of the industry. The cost per unit is less with cultured than with natural quartz because less labor and skill are required in orientation and cutting; little or no waste results from twinning and other defects; and the stones are grown to such size and shape that the minimum amount of quartz is wasted in cutting.

GROWTH OF CULTURED QUARTZ

Alpha quartz is grown from water solution. It is not possible to grow crystals of alpha quartz from the melt as is done with germanium, silicon, and other crystals. The melting point of SiO_2 is above 1700°C. Above 573°C, which is called the Curie temperature, SiO_2 exists as beta quartz, which has a hexagonal instead of a trigonal structure. When a crystal of beta quartz is cooled through the Curie temperature to become alpha quartz, very large thermal stresses occur and the crystal often shatters. Even if it does not break, the crystal usually becomes electrically twinned.

Cultured quartz is grown in large, very strong autoclaves which are typically 3 to 5 m long and half a meter or more in diameter with walls 10 to 15 cm thick. They are mounted, usually underground, in a vertical position and surrounded with heating coils and thermal insulation. The electric power to the heating elements is controlled by a computer programmed to establish the proper thermal gradients in the autoclave and to change them as the growth proceeds.

The seed crystals are mounted in racks somewhat like the rungs of a ladder. Pure silica, in chunks having a volume of a few cubic centimeters, provides the nutrient for the crystals. It is placed in the bottom of the autoclave, which is then filled nearly full of water to which is added some alkaline material such as sodium carbonate. Naturally the purest materials are required as is great care with cleanliness. The quality of the product is critically dependent on these factors as well as on the growing conditions.

The autoclave is sealed and the temperature is raised to approximately 350°C. The liquid expands until it fills the autoclave and exerts pressures of the order of 1000 atm on its walls. At these temperatures the solubility of SiO_2 is quite appreciable and, as in the usual case, increases with the temperature.

The lower end of the autoclave where the nutrient is placed is maintained at a higher temperature than the upper end where the seed crystals are located. The solution, which is saturated at the higher temperature, rises by convection to the cooler region, where it becomes supersaturated and SiO_2 is deposited on the seed crystals. Consequently, crystalline quartz grows at the expense of the pure silica nutrient.

The crystal grows most rapidly in the Z-direction where the rate is typically 0.5 mm/day. Thus a Y-bar of cultured quartz having dimensions 2 × 3 × 15 cm can be grown in about 1 month. Figure 2.5 shows a typical Y-bar of cultured quartz.

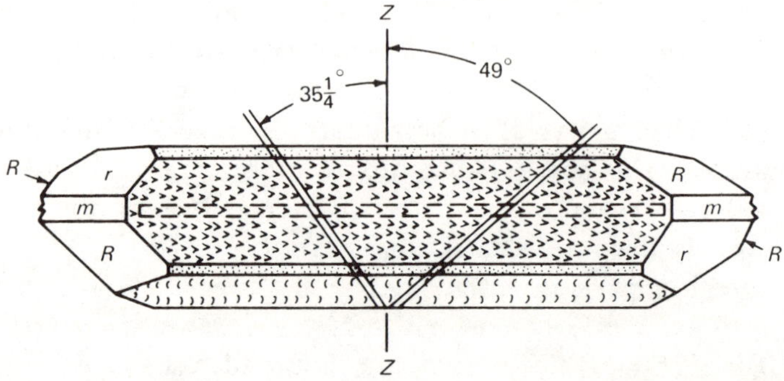

Fig. 2.5. Sketch of a specimen of cultured quartz grown in *Y*-bar form for use in fabricating AT- and BT-cut piezoids. The specimen is right-hand quartz with the positive end of the *X*-axis directed upward. The position of the seed crystal is indicated by the dotted lines. Compare the natural faces with those of Fig. 2.1. The lines sloping left from the *Z*-axis indicate the saw-cut position for cutting AT-cut plates. The lines sloping right are for the BT-cut.

ORIENTATION

Stones to be fabricated into AT- and BT-cut plates are usually grown in the form of *Y*-bars. Typically the seed crystal consists of a bar of quartz having a length of about 15 cm in the *Y*-direction, a width of 5 to 10 mm in the *X*-direction, and a thickness of 2 to 3 mm in the *Z*-direction. The seeds are usually cut from right-hand quartz, although this is of no particular importance except convenience. The final crystal has the form of a long bar with its maximum dimension in the *Y*-direction and its minimum dimension in the *X*-direction. The seed crystals are carefully selected to avoid twinning and other defects which could be propagated through the crystal as it grows.

Only two steps are required to obtain an accurately oriented AT- or BT-cut wafer using cultured quartz. This is very much simpler than the corresponding process using natural quartz, especially if most or all of the natural faces are absent. The first step is to grind a flat surface on the bar perpendicular to the *X*-axis. The bar is then mounted on this surface with the *X*-axis in a vertical position. The second step is to set the saw to cut parallel to the *X*-axis and at the correct angle with respect to the *Z*-axis. The correct angular position is achieved by the use of x-ray diffraction. The (10.1) planes which are parallel to the minor rhomb face (*r*) make an angle of 38° 13′

with the *Z*-axis. The AT-cut makes an angle of approximately 35°15′ with the *Z*-axis. Therefore, the dihedral angle between the (10.1) planes and the AT-cut is about 2°58′ (the exact value depending on the frequency and temperature range over which the unit is to be used). This is the angle which is measured by x-ray diffraction.

The BT-cut makes an angle of about −49° with the *Z*-axis. The rotation is in the opposite sense from that of the AT-cut. The (20.3) planes make an angle of 49°44′ with the *Z*-axis and are commonly used for orienting the BT-cut.

DEFECTS IN QUARTZ

Although cultured quartz has largely replaced natural quartz in the piezoelectric industry, the latter is still preferred for a few applications where the highest performance is required. Significant differences do exist between natural and cultural quartz and, indeed, between samples of natural quartz, although very sophisticated techniques are required to demonstrate them. These differences are associated with crystalline imperfections which are primarily impurity atoms, interstitial atoms, and lattice vacancies.

It is well known that some specimens of natural quartz are smoky or straw-colored. It has long been known, too, that some specimens of natural quartz are darkened by exposure to x-rays or gamma radiation. The development of smokiness is accompanied by changes in the elastic constants resulting in changes in the resonant frequencies which, in BT-cut plates, may be as great as 0.02 percent. At one time this change of frequency was considered as a method of adjusting the frequency of quartz crystal units. The idea was soon abandoned because some samples of quartz do not darken nor do their elastic constants change with exposure to radiation. Furthermore, the smokiness and frequency changes are both reversible by heating the quartz to a temperature of about 200°C. If the quartz is heated in a dark room, a flash of light may be observed as the temperature passes through the temperature at which the smokiness disappears. Although this phenomenon, called thermoluminescence, is well understood and is the basis of a dosimeter used to monitor exposure to radiation, it would take us too far into the physics of solids to discuss it further. It may be said, however, that crystalline defects are responsible for the effect.

Some impurities may be removed from quartz by placing the quartz in a strong electric field at elevated temperatures. Quartz which has been treated in this way is called *swept* quartz, since the impurity ions have been swept out by the electric field. Swept quartz is used in some very critical applications.

It seems likely that natural quartz grew at relatively low temperatures over very long periods of time. Cultured quartz, on the other hand, is grown at elevated temperatures in short periods of time. It is unavoidable that the number of interstitial atoms and lattice dislocations must increase as the growing temperature is increased. Consequently, the quality of cultured quartz is somewhat dependent upon the conditions under which it is grown.

The differences of growth conditions are reflected in the quality factor Q of the resonators made from the quartz. The presence of certain imperfections which produce absorption bands in the infrared absorption spectrum of the crystal have been correlated with the degradation of the Q of the resonator. Consequently infrared measurements are sometimes used to select quartz of superior quality for use in the most critical applications.

SYMMETRY RELATIONS IN CRYSTALS

Some properties of crystals are not unlike the corresponding properties in noncrystalline materials; for instance, density and specific heat. Such properties are called scalar quantities. Most physical properties, however, are dependent upon direction in the crystal. These are called tensor quantities.

The speed of propagation of a sound wave in a glass rod is independent of the direction in which the rod was cut from the original piece of glass. However, the speed of sound in a similar rod cut from a crystal of quartz is dependent upon the orientation of the rod relative to the crystallographic axes of the quartz. Thus, whereas a given physical property has only one value in a noncrystalline or amorphous material, it may have two or more in a crystalline material. It might be thought that the values of the coefficients in different directions would all be independent of each other, but they are not. Symmetry conditions in the crystal impose definite restrictions on both the number of independent coefficients and their values.

As an example of the restrictions imposed by symmetry conditions, we may consider the index of refraction of light in quartz. The index of refraction is the ratio of the velocity of light in vacuum to the velocity in the medium. A ray of yellow light traveling in the Z-direction in quartz travels at such a velocity that the index of refraction is 1.544. The index of refraction for light of the same color traveling in a direction perpendicular to the Z-direction is 1.553. A crystal, such as quartz, which has two indices of refraction, is said to be birefringent.

In the same way, the dielectric constant of quartz is different for fields applied parallel and perpendicular to the Z-axis. The value for fields in the Z- direction $k_{\|} = 4.60$, whereas the value for fields in the perpendicular direction $k_{\perp} = 4.51$.

The peculiar nature of crystals requires a special type of mathematics for use in solving problems involving their anisotropic properties. Matrix algebra is the mathematical tool best suited for the purpose. Since it is necessary to make some use of matrix algebra in designing quartz crystal units, a brief introduction to this very useful branch of mathematics will be given. A more complete treatment with further applications to problems of piezoid design has been given by Bond.[2]

THE ELEMENT OF MATRIX ALGEBRA

Matrix algebra is a branch of mathematics which is especially useful for handling sets of linear algebraic equations. Consider, for example, the set of linear equations

$$\begin{aligned} y_1 &= a_{11}x_1 + a_{12}x_2 + a_{13}x_3 \\ y_2 &= a_{21}x_1 + a_{22}x_2 + a_{23}x_3 \\ y_3 &= a_{31}x_1 + a_{32}x_2 + a_{33}x_3 \end{aligned} \tag{1}$$

[2] Walter L. Bond, "The Mathematics of the Physical Properties of Crystals," *B.S.T.J.*, 22: 1-72 (1943).

in which the y's are variables and the a_{ij}'s are constant coefficients. The equation above is written in matrix notation as

$$y = ax \tag{2}$$

where y is written for the array $\begin{pmatrix} y_1 \\ y_2 \\ y_3 \end{pmatrix}$, x for the array $\begin{pmatrix} x_1 \\ x_2 \\ x_3 \end{pmatrix}$

and a for the matrix $\begin{pmatrix} a_{11} & a_{12} & a_{13} \\ a_{21} & a_{22} & a_{23} \\ a_{31} & a_{32} & a_{33} \end{pmatrix}$

Expression (2) is made formally equal to expression (1) by defining an operation called matrix multiplication. Written out in full, (2)

becomes $$\begin{pmatrix} y_1 \\ y_2 \\ y_3 \end{pmatrix} \begin{matrix} = \\ = \\ = \end{matrix} \begin{pmatrix} a_{11} & a_{12} & a_{13} \\ a_{21} & a_{22} & a_{23} \\ a_{31} & a_{32} & a_{33} \end{pmatrix} \begin{pmatrix} x_1 \\ x_2 \\ x_3 \end{pmatrix} \tag{3}$$

Matrix multiplication consists of forming the sum of the products obtained by multiplying the terms in the rows of a by the corresponding terms in the column of x. Mechanically this is readily done by moving a finger of the left hand across the rows of a while a finger of the right hand moves down the column of x performing the multiplication at each step and accumulating the sum mentally. Computer programs are readily prepared for performing the operation. By defining multiplication in this way, (3) is made formally equivalent to (1).

The value of matrix algebra is due to the ease with which it permits handling large sets of simultaneous equations. For example, it can be used to "solve" sets of simultaneous equations of the following type.

Let Eq. (1) above be represented by

$$y = ax \tag{4}$$

and let

$$x = bz \tag{5}$$

represent the set
$$\begin{aligned} x_1 &= b_{11}z_1 + b_{12}z_2 + b_{13}z_3 \\ x_2 &= b_{21}z_1 + b_{22}z_2 + b_{23}z_3 \\ x_3 &= b_{31}z_1 + b_{32}z_2 + b_{33}z_3 \end{aligned} \tag{6}$$

The x's may be eliminated and the y's expressed in terms of the z's by writing formally

$$y = abz \tag{7}$$

where bz is substituted for x in (4). It must be noted that the substitution must be made in the particular order, for ab is not equivalent to ba. Matrix multiplication is *not* commutative.

Written out in full, (7) becomes

$$\begin{pmatrix} y_1 \\ y_2 \\ y_3 \end{pmatrix} = \begin{pmatrix} a_{11} & a_{12} & a_{13} \\ a_{21} & a_{22} & a_{23} \\ a_{31} & a_{32} & a_{33} \end{pmatrix} \begin{pmatrix} b_{11} & b_{12} & b_{13} \\ b_{21} & b_{22} & b_{23} \\ b_{31} & b_{32} & b_{33} \end{pmatrix} \begin{pmatrix} z_1 \\ z_2 \\ z_3 \end{pmatrix} \tag{8}$$

In "solving" (8), the product ab is first formed. The product of ab is a square matrix, the ij term of which is found by forming the product of the ith row of a with the jth column of b. Carrying out the multiplication in this way results in the matrix

$$ab =$$

$$\begin{pmatrix} a_{11}b_{11}+a_{12}b_{21}+a_{13}b_{31} & a_{11}b_{12}+a_{12}b_{22}+a_{13}b_{32} & a_{11}b_{13}+a_{12}b_{23}+a_{13}b_{33} \\ a_{21}b_{11}+a_{22}b_{21}+a_{23}b_{31} & a_{21}b_{12}+a_{22}b_{22}+a_{23}b_{32} & a_{21}b_{13}+a_{22}b_{23}+a_{23}b_{33} \\ a_{31}b_{11}+a_{32}b_{21}+a_{33}b_{31} & a_{31}b_{12}+a_{32}b_{22}+a_{33}b_{32} & a_{31}b_{13}+a_{32}b_{23}+a_{33}b_{33} \end{pmatrix}$$

Finally the product ab is multiplied into z giving

$$\begin{aligned} y_1 &= (a_{11}b_{11} + a_{12}b_{21} + a_{13}b_{31})z_1 + (a_{11}b_{12} + a_{12}b_{22} + a_{13}b_{32})z_2 + \\ &\quad (a_{11}b_{13} + a_{12}b_{23} + a_{13}b_{33})z_3 \\ y_2 &= (a_{21}b_{11} + a_{22}b_{21} + a_{23}b_{31})z_1 + (a_{21}b_{12} + a_{22}b_{22} + a_{23}b_{32})z_2 + \\ &\quad (a_{21}b_{13} + a_{22}b_{23} + a_{23}b_{33})z_3 \\ y_3 &= (a_{31}b_{11} + a_{32}b_{21} + a_{33}b_{31})z_1 + (a_{31}b_{12} + a_{32}b_{22} + a_{33}b_{32})z_2 + \\ &\quad (a_{31}b_{13} + a_{32}b_{23} + a_{33}b_{33})z_3 \end{aligned}$$

In the last set of equations the x's have been eliminated and the y's are expressed in terms of the z's.

A square matrix is one having equal numbers of rows and columns. In a square matrix the diagonal from upper left to lower right is called the major diagonal. A matrix consisting of 1's on the major diagonal with all other terms zero is called the identity matrix I. Multiplying any matrix by I leaves the matrix unchanged. I in matrix algebra is therefore analogous to unity in ordinary algebra. The reciprocal matrix A^{-1} is defined by

$$A A^{-1} = I$$

The matrix equation $Y = AX$ can be considered to be solved for X by writing

$$A^{-1} Y = A^{-1} A X = IX = X$$

Rules for computing the reciprocal matrix may be found in books on the subject of matrix algebra. A simple method involves the following four steps.

1. Interchange rows and columns.
2. Replace each term by the determinant of its minor. (The minor of a given term is the remainder of the matrix after the row and column containing the given term have been removed.)
3. Multiply alternate terms by 1 and −1 starting with 1 in the upper-left-hand-corner.
4. Divide each term by the determinant of the original matrix.

To illustrate these steps, we compute the reciprocal of the matrix A where

$$A = \begin{pmatrix} 1 & 2 & 3 \\ 4 & 5 & 6 \\ 2 & 8 & 4 \end{pmatrix}$$

The application of step 1 gives

$$\begin{pmatrix} 1 & 4 & 2 \\ 2 & 5 & 8 \\ 3 & 6 & 4 \end{pmatrix}$$

Step 2 gives

$$\begin{pmatrix} -28 & -16 & -3 \\ 4 & -2 & -6 \\ 22 & 4 & -3 \end{pmatrix}$$

Step 3 gives

$$\begin{pmatrix} -28 & 16 & -3 \\ -4 & -2 & 6 \\ 22 & -4 & -3 \end{pmatrix}$$

Finally, using step 4,

$$A^{-1} = \begin{pmatrix} -28/30 & 16/30 & -3/30 \\ -4/30 & -2/30 & 6/30 \\ 22/30 & -4/30 & -3/30 \end{pmatrix}$$

It can readily be shown that

$$A^{-1}A = \begin{pmatrix} 1 & 0 & 0 \\ 0 & 1 & 0 \\ 0 & 0 & 1 \end{pmatrix} = I$$

The reciprocal matrix can be considered to be the solution of the original matrix, for if $I = AX$, then $X = A^{-1}I$. It follows therefore that the reciprocal matrix is defined only for a square matrix. The calculation of the reciprocal of a large matrix can be quite tedious, but computer programs are available for computing A^{-1} for any matrix.

ROTATION OF AXES

One of the principle applications of matrix algebra to the physics of crystals is that of determining the values of the coefficients of the physical properties for any given direction in the crystal. This is especially important in the design of piezoids, which must often be oriented in such a way that the directions of the electric fields and wave propagation are not parallel to the crystallographic axes. With the aid of matrix algebra it is possible to compute the value of any coefficient provided only that the corresponding values are known for the directions of the crystallographic axes.

To illustrate how this is done, we first show how to find the components of a vector in a new coordinate system given the components in the original system. Let the vector **V** have the components V_x, V_y, V_z in the right-hand coordinate system XYZ. Then let the coordinate system be rotated about the X axis by an angle θ forming the new coordinate system $XY'Z'$.

It is easy to see from Fig. 2.6 that the components of the vector V' in the new system expressed in terms of the components of V in the old system are

$$\begin{aligned} V_x' &= V_x \\ V_y' &= V_y \cos\theta + V_z \sin\theta \\ V_z' &= -V_y \sin\theta + V_z \cos\theta \end{aligned} \tag{9}$$

In matrix notation the set of equations above may be written as

$$V' = aV$$

where a is the matrix

$$a = \begin{pmatrix} 1 & 0 & 0 \\ 0 & c & s \\ 0 & -s & c \end{pmatrix} \quad \text{in which} \quad \begin{aligned} c &= \cos\theta \\ s &= \sin\theta \end{aligned} \tag{10a}$$

The matrix a may be thought of as performing a rotation of the vector **V** by an angle θ about the X axis.

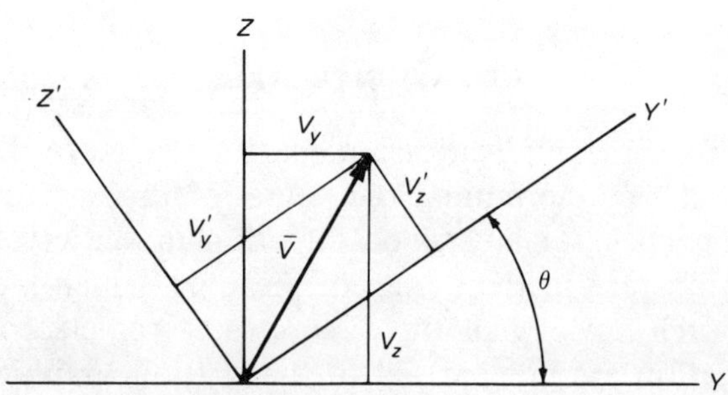

Fig. 2.6. Coordinate system *XYZ* rotated by an angle θ to form a new coordinate system *XY'Z'*.

In the same way the matrix

$$b = \begin{pmatrix} c & 0 & -s \\ 0 & 1 & 0 \\ s & 0 & c \end{pmatrix} \quad \text{in which} \quad \begin{matrix} c = \cos\psi \\ s = \sin\psi \end{matrix} \tag{10b}$$

performs a rotation of angle ψ about the Y axis. The matrix

$$c = \begin{pmatrix} c & s & 0 \\ -s & c & 0 \\ 0 & 0 & 1 \end{pmatrix} \quad \text{in which} \quad \begin{matrix} c = \cos\phi \\ s = \sin\phi \end{matrix} \tag{10c}$$

performs a rotation of ϕ about the Z axis.

To illustrate the application of the method we may consider any vector property of the crystal such as electric induction, magnetic induction, or thermal or electric conductivity. We use the first to illustrate the procedure.

In an isotropic material the electric induction **D** is related to the electric-field intensity **E** by the equation **D** = k **E**, where k is called the dielectric constant. **D** and **E** are parallel. In an anisotropic material, such as quartz, **D** and **E** are not necessarily parallel and in general each component of **D** depends on all three components of **E**. We therefore have for the components of the induction **D** in terms of the components of the electric field **E**,

$$\begin{aligned} D_x &= k_{11}E_x + k_{12}E_y + k_{13}E_z \\ D_y &= k_{21}E_x + k_{22}E_y + k_{23}E_z \\ D_z &= k_{31}E_x + k_{32}E_y + k_{33}E_z \end{aligned} \tag{11}$$

or in matrix notation

$$D = kE$$

Both **D** and **E** being vectors, are rotated according to the rules developed above, so that for a rotation of axes by the matrix a we have

$$D' = aD \qquad \text{and} \qquad E' = aE \qquad \text{and} \qquad a^{-1}E' = a^{-1}aE = E$$

$$D' = a\,k\,E = a\,k\,a^{-1}\,E'$$

so that $D' = k'E'$ where $k' = a\,k\,a^{-1}$

The convenience and brevity of the notation above should not be permitted to obscure the physical picture of the operation.

A very important and powerful application of the rotation of axes is involved in investigating the effects of crystal symmetry. For example, we will show that of the nine possible terms in the matrix for the dielectric constant of quartz, all but three are zero and two of these are equal. The same result can be obtained for any other vector property.

The Z-axis in quartz is, as we have seen, an axis of three fold symmetry. A rotation of 120° about the Z-axis must, therefore, leave all the physical properties unchanged. As an example we will perform such a rotation on the dielectric matrix [Eq. (11)]. For a rotation of 120° about the Z-axis, we have from Eq. (10c),

$$a = \begin{pmatrix} -1/2 & \sqrt{3}/2 & 0 \\ -\sqrt{3}/2 & -1/2 & 0 \\ 0 & 0 & 1 \end{pmatrix} \quad \text{and} \quad a^{-1} = \begin{pmatrix} -1/2 & -\sqrt{3}/2 & 0 \\ \sqrt{3}/2 & -1/2 & 0 \\ 0 & 0 & 1 \end{pmatrix}$$

The k matrix is, of course,

$$k = \begin{pmatrix} k_{11} & k_{12} & k_{13} \\ k_{21} & k_{22} & k_{23} \\ k_{31} & k_{32} & k_{33} \end{pmatrix}$$

the $k' = aka^{-1}$ matrix for a rotation of 120° about the Z axis is then

$$k' = \begin{pmatrix} \begin{matrix} 1/4\,k_{11} - \sqrt{3}/4\,k_{21} \\ +\sqrt{3}/4\,k_{12} + 3/4\,k_{22} \end{matrix} & \begin{matrix} \sqrt{3}/4\,k_{11} - 3/4\,k_{21} \\ +1/4\,k_{12} - \sqrt{3}/4\,k_{22} \end{matrix} & -1/2\,k_{13} + \sqrt{3}/2\,k_{23} \\ \begin{matrix} \sqrt{3}/4\,k_{11} + 1/4\,k_{21} \\ -3/4\,k_{12} - \sqrt{3}/4\,k_{22} \end{matrix} & \begin{matrix} 3/4\,k_{11} + \sqrt{3}/4\,k_{21} \\ +\sqrt{3}/4\,k_{12} + 1/4\,k_{22} \end{matrix} & -\sqrt{3}/2\,k_{13} - 1/2\,k_{23} \\ -1/2\,k_{31} + \sqrt{3}/2\,k_{32} & -\sqrt{3}/2\,k_{31} - 1/2\,k_{32} & k_{33} \end{pmatrix}$$

Since any rotation which is compatible with symmetry conditions must leave the values of the coefficients unchanged, it follows that

k_{11} must be equal to k'_{11}, $k_{12} = k'_{12}$, etc. These relations lead to a set of nine simultaneous equations which are satisfied if and only if

$$k_{21} = k_{13} = k_{31} = k_{12} = k_{23} = k_{32} = 0$$

and $$k_{11} = k_{22} \qquad k'_{33} = k_{33}$$

We thus find that quartz has only two independent dielectric constants, k_{11} and k_{33}, i.e., the values perpendicular and parallel, respectively, to the Z-axis. To prove that k_{33} is indeed not equal to k_{11} it is necessary to perform another rotation of 180° about either the X- or Y-axis. This operation, which is also compatible with the symmetry of quartz, is carried out in the same way and leads to the result that the dielectric matrix for quartz is

$$k = \begin{pmatrix} k_{11} & 0 & 0 \\ 0 & k_{11} & 0 \\ 0 & 0 & k_{33} \end{pmatrix}$$

The dielectric matrix for quartz thus contains only three nonzero terms and two of these are equal. A similar matrix describes the thermal and electric conductivities and all other vector properties of quartz. The fact that many of the possible coefficients are zero is a consequence of the symmetry conditions in the crystal and fortunate for the individual who finds it necessary to make computations which involve the physical properties of crystalline materials.

THE DIELECTRIC CONSTANT AND THERMAL EXPANSION

As illustrations of the use of the matrix algebra we will compute the dielectric constant and thermal expansion coefficient for an arbitrary direction in the quartz crystal. To do so we need only compute the values in the YZ-plane, since the value of any vector quantity is constant in the XY-plane in quartz.

We will determine the dielectric constant in the thickness direction of a quartz plate whose normal makes an angle θ with the Z-axis by performing a rotation about the X-axis. We have already seen that $D' = k'E'$, where $k' = a\,k\,a^{-1}$. The matrix which performs a rotation about the X-axis is given by Eq. (10a). Its reciprocal can be computed by the recipe given on page 28. The required operation is therefore

$$k' = \begin{pmatrix} 1 & 0 & 0 \\ 0 & c & s \\ 0 & -s & c \end{pmatrix} \begin{pmatrix} k_{11} & 0 & 0 \\ 0 & k_{11} & 0 \\ 0 & 0 & k_{33} \end{pmatrix} \begin{pmatrix} 1 & 0 & 0 \\ 0 & c & -s \\ 0 & s & c \end{pmatrix}$$

where $c = \cos\theta$ and $s = \sin\theta$.

Carrying out the indicated operations gives

$$k' = \begin{pmatrix} k_{11} & 0 & 0 \\ 0 & k_{11}c^2 + k_{33}s^2 & (k_{33} - k_{11})sc \\ 0 & (k_{33} - k_{11})sc & k_{11}s^2 + k_{33}c^2 \end{pmatrix}$$

If $\theta = 0$ the matrix k' reduces to the matrix k and if $\theta = 90°$, the result is to interchange the Y- and Z-axes. The meaning is obvious geometrically.

In order to interpret the significance of the terms in the k' matrix, it is necessary to recall the significance of a matrix which is simply a table of the coefficients of the terms in a set of linear equations. The term k_{ij} relates the induction in the i direction with the electric field in the j direction. The terms $k'_{12} = k'_{21} = k'_{13} = k'_{31} = 0$ imply that no electric displacement is induced in the X'-direction by components of an electric field in the Y'- and Z'-directions and that no electric displacement in the Y'- or Z'-directions is produced by a field in the X'-direction.

The term $k'_{22} = k_{11} \cos^2\theta + k_{33} \sin^2\theta$ indicates that the electric displacement in the Y'-direction depends on both k_{11} and k_{33} as well as the angle θ. The same is true of k'_{33}.

The terms $k'_{23} = k'_{32} = (k_{33} - k_{11}) \sin\theta \cos\theta$, which is not always zero, implies that a field in the Y'-direction may produce a displacement in the Z'-direction, and vice versa. It is a bit surprising at first to find a component of electric displacement at right angles to the electric field which produces the displacement. This, of course, does not occur in isotropic materials. But in a crystalline dielectric the resultant of the displacement may not be parallel to the electric field;

it may be said to "follow the grain of the crystal" somewhat as a nail tends to follow the grain of the wood.

We now compute the effective dielectric constant for an AT-cut plate in which the plate normal makes an angle of (approximately) 55° with the Z-axis. The component of the dielectric constant k' in the direction normal to the plate is given by

$$k'_{33} = k_{11} \sin^2 55° + k_{33} \cos^2 55°$$

Since[3] $k_{11} = 4.51$ and $k_{33} = 4.60$, the value of the effective dielectric constant for the AT-cut plate is

$$k'_{33} = 4.54$$

Later we will have occasion to use this value in calculating the static capacitance of the crystal unit.

It is of interest to calculate the value of k'_{32}, which yields the component of the displacement in the Y'-direction. It has the value

$$\begin{aligned} k'_{32} &= (k_{33} - k_{11}) \sin 55° \cos 55° \\ &= 0.042 \end{aligned}$$

The Z'-component of the displacement is large compared with the Y'-component because the values of k_{11} and k_{33} are so nearly equal. Consequently, the orthogonal component may usually be disregarded in computing the effective dielectric constant in quartz. This may not be possible in other crystals in which the values are much different. In the example of the AT-cut, the displacement makes an angle arctan (0.042/4.54), which is about 0.5°, with the direction of the electric field.

Exactly the same procedure is followed in calculating other vector-vector properties such as the electrical and thermal conductivities. Again it is found that the direction of current flow is not necessarily

[3] The dielectric constants of alpha quartz at 20°C are $k_{33} = k_{\parallel} = 4.60 \pm 0.01$ and $k_{11} = k_{\perp} = 4.51 \pm 0.01$. Virgil E. Bottom, "Dielectric Constants of Quartz," *J. Appl. Phys.*, **43** (4): 1493–1495 (1972).

parallel to the applied field nor is the direction of heat flow necessarily parallel to the temperature gradient.

The linear coefficient of thermal expansion for an arbitrary direction in a crystal is calculated in a slightly different way, since temperature is a scalar quantity. By a proper choice of axes (which in quartz is the usual *XYZ* axial system) the linear coefficients of thermal expansion can be expressed in the form

$$\alpha = \begin{pmatrix} \alpha_{11} & 0 & 0 \\ 0 & \alpha_{22} & 0 \\ 0 & 0 & \alpha_{33} \end{pmatrix}$$

The change Δ **L** in a vector **L** resulting from unit temperature change (Fig. 2.7) may be expressed in matrix notation by

$$\Delta L = \alpha L$$

Since both **L** and Δ **L** are vectors they transform as

$$L' = a L \qquad \text{and} \qquad \Delta L' = a\,\Delta L$$

where[4]

$$a = (\cos\phi \sin\theta, \sin\phi \sin\theta, \cos\theta)$$

is the matrix of the coordinates of the vector L. Hence $\Delta L' = \alpha' L'$, where $\alpha' = a\,\alpha\,a^{-1}$.

Carrying out the indicated operation,

$$\begin{pmatrix} \cos\phi \sin\theta, \sin\phi \sin\theta, \cos\theta \end{pmatrix} \begin{pmatrix} \alpha_{11} & 0 & 0 \\ 0 & \alpha_{22} & 0 \\ 0 & 0 & \alpha_{33} \end{pmatrix} \begin{pmatrix} \cos\phi \sin\theta \\ \sin\phi \sin\theta \\ \cos\theta \end{pmatrix}$$

we have

$$\alpha'_{\theta,\phi} = \alpha_{11} \cos^2\phi \sin^2\theta + \alpha_{22} \sin^2\phi \sin^2\theta + \alpha_{33} \cos^2\theta$$

[4] The IEEE (or IRE) notation is used in the equations above, whereas the Bond notation is used in Eqs. (87) and (92). This accounts for the difference in the value of θ for the AT-cut. See Fig. 8.1.

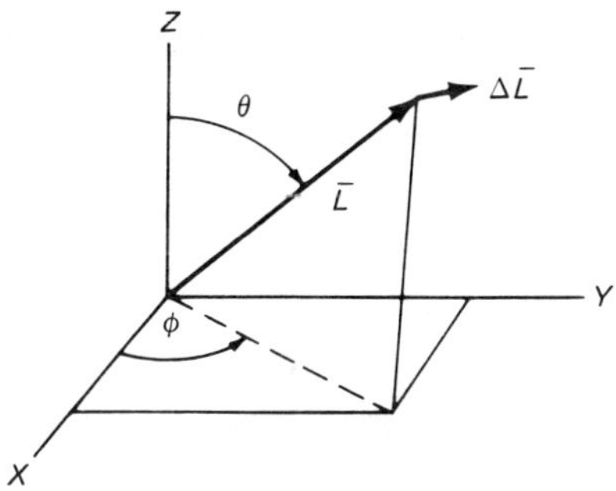

Fig. 2.7. The effect of thermal expansion is to change L to L + Δ L.

In quartz $\alpha_{11} = \alpha_{22}$ so the expression above reduces to

$$\alpha'_{\theta} = \alpha_{11} \sin^2 \theta + \alpha_{33} \cos^2 \theta$$

In the case of the AT-cut plate in which $\theta = 55°$ (see footnote 4), the linear coefficient of thermal expansion in the thickness direction is

$$\alpha_{\theta} = 55° = (14.3 \sin^2 55° + 7.8 \cos^2 55°) \times 10^{-6} \text{ per °C}$$
$$= 12.2 \times 10^{-6} \text{ per °C}$$

3
STRESS AND STRAIN RELATIONSHIPS IN QUARTZ

When a solid is subjected to forces which change or tend to change its shape or the distance between elements of the body, the body is said to be subjected to stress. The actual stresses which occur in solids are often quite complex, but it can be shown that the most complex stress is a combination of compressional, tensile, and shear stresses. A compressional stress tends to change the volume of a body. Tension tends to change the length. A shear stress tends to change the shape of the body. We will have no need to consider the effect of pressure, since few piezoelectric devices are produced which operate by subjecting a crystal to hydrostatic pressure. However, both tensile and shear stresses are common in piezoids and a clear understanding of their nature is essential to the design and operation of the quartz crystal unit.

DEFINITIONS OF STRESS AND STRAIN

A stress is defined as a force per unit area and therefore has the dimensions of pounds per square inch, newtons per square meter, dynes per square centimeter, etc. A tensile stress is the value of the tensile force divided by the area to which the force is applied. A shear stress is also a force divided by the area over which it is applied. In the case of a tensile stress the force is applied in a direction perpendicular to the area, while in a shear stress the direction of the force is parallel to the plane of the area. The two types of stresses are illustrated in Figure 3.1,

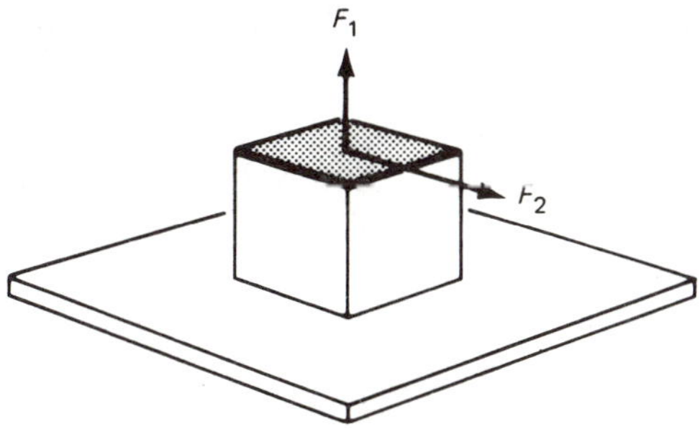

Fig. 3.1. F_1 **produces a longitudinal or tensile stress;** F_2 **produces a shear stress.**

where a solid body in the form of a cube is attached to a fixed surface and has a rigid plate attached to its upper surface. The area of the upper surface and of the rigid plate is A. The force F_1, perpendicular to A, produces a tensile stress F_1/A. The force F_2, parallel to A, produces a shear stress F_2/A.

Strain is the deformation associated with stress. Sometimes it is convenient to consider strain to be the result of a stress, but mathematically, as well as physically, it is quite logical to consider either as the cause and the other as the effect. The measure of a strain is the degree of the deformation. For instance, if the deformation consists of a change of length due to a tensile stress, the strain is the change in length divided by the initial length. The strain is therefore a dimensionless quantity.

In Fig. 3.2*a* the tensile stress F_1/A has resulted in an elongation of the dimension OQ by an amount QQ'. The longitudinal strain is thus given by QQ'/OQ. Some quartz piezoids vibrate in this way.

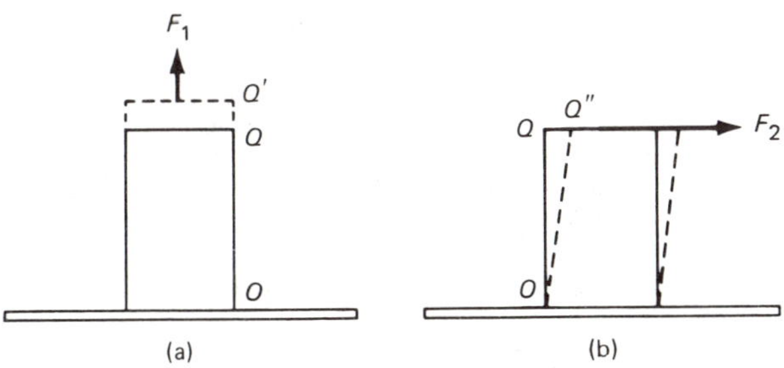

Fig. 3.2. (*a*) Longitudinal strain and (*b*) shear strain.

Figure 3.2*b* shows a shear strain. The shear stress F_2/A has produced a displacement QQ'' and a shear strain QQ''/OQ. Shear strains, like longitudinal strains, are very small in quartz, the maximum value being set by the elastic limit.[1] Since the shear displacement QQ'' is very small, it is customary to speak of the angle of shear and to define the angle (in radians) as QQ''/OQ.

Modes of vibration employing shear stresses and strains are used in many quartz piezoids. It is necessary, therefore, to expand upon the concept of shear stress and strain. A complete treatment of this topic is quite involved, but for our purpose it will be sufficient to distinguish between a simple and a pure shear and to note the type of displacement involved in each.

The concept of a simple shear can be understood from Fig. 3.3*a* in which the shaded area $OABC$ represents a square unstrained area. A simple shear strain X_y results in the deformation of $OABC$ into the shape $OAB'C'$. The simple shear Y_x distorts $OABC$ into the shape $OA''B''C$ as shown in *b*. Thus a simple shear strain is produced by holding one side of the square fixed while distorting it into a rhombus. A simple shear can be transformed into a pure shear by a change of coordinates, i.e., the deformation of the body is the same in both

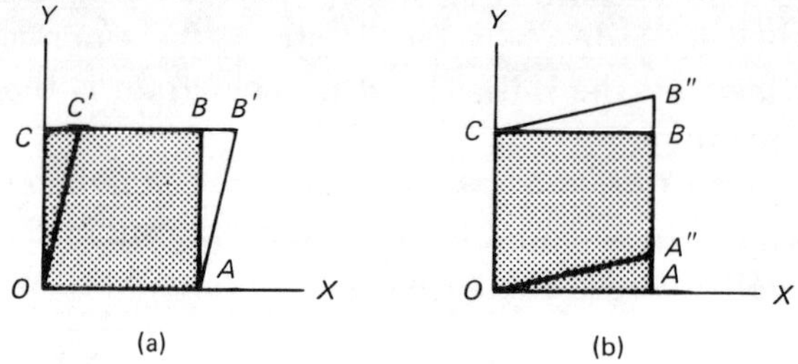

Fig. 3.3. Simple shear strains.

[1] The elastic limit, or breaking strain, in a crystal such as quartz is very hard to determine. The measured elastic limit, or fracture point, is far smaller than the theoretical value in quartz and other high-purity crystals such as silicon and germanium. It is believed that these crystals fail by fracture caused by stress concentration at minute surface cracks. The stress-strain ratio in quartz appears to be linear up to the fracture point but, of course, not necessarily to the true elastic limit. Some phenomena which have been attributed to nonelastic properties of quartz are probably due to other effects.

In any case, the strains in piezoelectric devices used in the communication field should be very much less than the elastic limit.

simple and pure shears. Physically, the difference depends upon the manner in which the body is supported and the forces applied. In the ideal case of a thin square plate subjected to a shear strain in the plane of the plate, the strain may be regarded as a simple shear if the plate is clamped at one edge, as shown in Fig. 3.4*a*. Exactly the same deformation will occur if the plate is supported at its center while equal and opposite forces are applied, as shown in *b*. It should also be apparent that the same deformation can be produced by the application of compressive forces F_2 along one diagonal while extensional forces F_3 are applied along the other diagonal, as shown in *c*.

A pure shear is equivalent to two longitudinal strains at right angles to each other. This can be seen by considering the square *ABCD*, Fig. 3.5, marked on the plane surface S_1. When the surface is distorted by a shear strain in the same plane, the square becomes the rectangle $A'B'C'D'$. In the same way a circle is distorted into an ellipse. The equivalence of a pure shear strain to two extensional strains at right angles is important in the design of certain oscillator plates, especially those designated as *CT* and *DT*. These plates vibrate in such a way that the diagonals (or perpendicular diameters) become alternately longer and shorter. Such a mode of motion is called the *face-shear* mode.

It is customary, in discussing shear strains and stresses, to speak of the axis of stress or strain or of a "shear about a given axis." These terms refer to the direction normal to the plane in which the shear occurs. This is also the axis about which rotation occurs in a simple shear. The axis of a shear strain is analogous mathematically to the direction of an extensional strain.

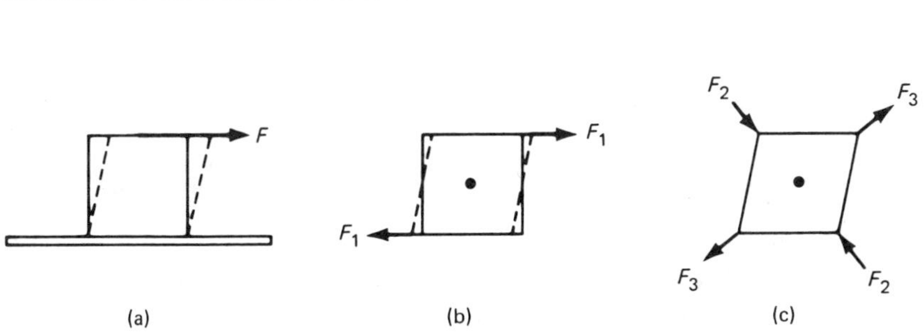

Fig. 3.4. Pure shear strain and its relation to simple shear strain.

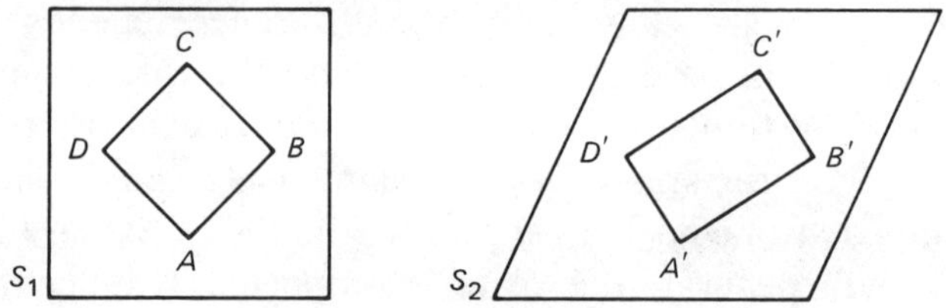

Fig. 3.5. The shear strain is equivalent to two orthogonal extensional strains.

DEFINITIONS OF THE STRESS-STRAIN RELATIONSHIPS

According to Hooke's law the strain which a solid body experiences is linearly proportional to the applied stress. Like many of the laws of physics, Hooke's law is an approximation which is valid only under certain conditions. A very well made spring "obeys" Hooke's law over a wide range of strain, but the law eventually fails as the spring reaches its maximum extension. A rubber band does not even approximately obey Hooke's law.

Hooke's law may be used to describe the stress-strain relationship in many solids provided the strains are small enough. If the stresses and strains are small enough that their ratio may be considered to be a constant, independent of the magnitude of the stress, we may define a quantity called the *modulus of elasticity*, or *elastic modulus*. The most-common example is Young's modulus, which is used to describe the elastic behavior of one-dimensional objects such as thin wires. The Young's modulus (YM) of a thin wire is given by

$$\text{YM} = \frac{F/A}{e/L} = \frac{F\,L}{A\,e}$$

where

A = area of cross section of wire
L = initial length
e = extension resulting from the application of force F

The dimensions of a modulus are those of force/area.

The modulus of shear is defined as the ratio of the shear stress, Fig. 3.2*b*, to the resulting strain defined as the angle of shear.

The bulk modulus is defined as the ratio of the hydrostatic pressure divided by the fractional change of volume. We have no need of the bulk modulus in this book but the concept of the shear modulus is most important.

In work in piezoelectricity it is more common to speak of the stress/strain ratio as a *stiffness coefficient*. The inverse ratio, strain/stress, is called a *compliance coefficient*. The term *constant* is often used in place of *coefficient* in both definitions, and the stiffness coefficients are commonly called the *elastic constants*.

THE GENERALIZED STRESS-STRAIN RELATIONSHIPS

For small displacements it can be shown that the most general strain in a solid can be resolved into a linear combination of an extensional and a pure shear strain. Any extensional strain can be resolved into three components parallel to a set of orthogonal axes. Likewise, any shear strain can also be resolved into three shear strains about the same three axes.

It is easily understood that a longitudinal stress in an isotropic solid will produce not only a longitudinal strain parallel to the stress but also lateral strains perpendicular to it. The ratio of the lateral to the longitudinal strain is called Poisson's ratio. It is not conceivable, however, that a tensile stress in an isotropic material should produce any shear strains. But it does !! See Fig 3.5

The situation in anisotropic (crystalline) materials is more complex. An extensional stress produces extensional strain and lateral contraction, and shear strains may also exist. For example, if a tensile stress is applied to a thin rod of quartz cut with its length parallel to the X-axis, not only does the rod become longer and thinner, but it also tends to rotate about its longitudinal axis.

It follows therefore that to completely describe the stress-strain relationships in crystalline materials, any component of stress must be considered to be capable of producing any component of strain. Since we have 6 components of stress and 6 of strain, 36 constants must be employed to describe the behavior in the general case. Fortunately, many of the constants are zero and simplifying relations exist among the others so that the number is greatly reduced. In quartz, for example, 18 of the 36 are zero and only 6 of the remainder are independent.

For a crystalline material the linear stress-strain relationship has the form:

$$\begin{aligned}
-x_x &= s_{11}X_x + s_{12}Y_y + s_{13}Z_z + s_{14}Y_z + s_{15}Z_x + s_{16}X_y \\
-y_y &= s_{21}X_x + s_{22}Y_y + s_{23}Z_z + s_{24}Y_z + s_{25}Z_x + s_{26}X_y \\
-z_z &= s_{31}X_x + s_{32}Y_y + s_{33}Z_z + s_{34}Y_z + s_{35}Z_x + s_{36}X_y \\
-y_z &= s_{41}X_x + s_{42}Y_y + s_{43}Z_z + s_{44}Y_z + s_{45}Z_x + s_{46}X_y \\
-z_x &= s_{51}X_x + s_{52}Y_y + s_{53}Z_z + s_{54}Y_z + s_{55}Z_x + s_{56}X_y \\
-x_y &= s_{61}X_x + s_{62}Y_y + s_{63}Z_z + s_{64}Y_z + s_{65}Z_x + s_{66}X_y
\end{aligned} \tag{12}$$

where x_x, y_y, and z_z are the extensional strain components parallel to the X, Y, and Z axes, respectively, and y_z, z_x, and x_y are the shear strain components about the X, Y, and Z axes, respectively. The negative signs result from the convention that a stress which tends to compress a body is called positive, while a decrease in the length is called a negative strain. Therefore, a positive stress is associated with a negative strain.

The terms X_x, Y_y, and Z_z refer to extensional stresses and the terms Y_z, Z_x, and X_y refer to the shear stresses. In a pure shear, as we have seen, X_y is mathematically equivalent to Y_x, thereby reducing the number of stress-strain components from 9 to 6 and the number of terms in the matrix from 81 to 36.

Sometimes it is desirable to express the stresses in terms of the strains. In this case we have the following set of equations:

$$\begin{aligned}
-X_x &= c_{11}x_x + c_{12}y_y + c_{13}z_z + c_{14}y_z + c_{15}z_x + c_{16}x_y \\
-Y_y &= c_{21}x_x + c_{22}y_y + c_{23}z_z + c_{24}y_z + c_{25}z_x + c_{26}x_y \\
-Z_z &= c_{31}x_x + c_{32}y_y + c_{33}z_z + c_{34}y_z + c_{35}z_x + c_{36}x_y \\
-Y_z &= c_{41}x_x + c_{42}y_y + c_{43}z_z + c_{44}y_z + c_{45}z_x + c_{46}x_y \\
-Z_x &= c_{51}x_x + c_{52}y_y + c_{53}z_z + c_{54}y_z + c_{55}z_x + c_{56}x_y \\
-X_y &= c_{61}x_x + c_{62}y_y + c_{63}z_z + c_{64}y_z + c_{65}z_z + c_{66}x_y
\end{aligned} \tag{13}$$

Equations (12) and (13) may be considered the generalized form of Hooke's law for crystalline materials.

In matrix notation the stress-strain relations, Eqs. (12) and (13), may be written, respectively,

$$-x = s\,X \tag{12a}$$

and

$$-X = c\,x \tag{13a}$$

where s is the square matrix

$$s = \begin{pmatrix} s_{11} & s_{12} & s_{13} & s_{14} & s_{15} & s_{16} \\ s_{21} & s_{22} & s_{23} & s_{24} & s_{25} & s_{26} \\ s_{31} & s_{32} & s_{33} & s_{34} & s_{35} & s_{36} \\ s_{41} & s_{42} & s_{43} & s_{44} & s_{45} & s_{46} \\ s_{51} & s_{52} & s_{53} & s_{54} & s_{55} & s_{56} \\ s_{61} & s_{62} & s_{63} & s_{64} & s_{65} & s_{66} \end{pmatrix}$$

The c matrix is entirely analogous.

The s's in Eq. (12) are called the *compliance* coefficients and the c's in Eq. (13) are called the *stiffness* coefficients, a choice of notation which obviously was not chosen for its mnemonic value.

The principle of conservation of energy requires that $c_{ij} = c_{ji}$ and that $s_{ij} = s_{ji}$, thus reducing the number of independent coefficients in each set from 36 to 21. The symmetry conditions require that many of the coefficients be zero and that simple relationships exist between others. In a cubic crystal only three independent elastic coefficients exist, but in quartz the number of nonzero, independent coefficients is six.

By performing rotations of the c and s matrices as permitted by the holoaxial, trigonal symmetry of quartz, it may be shown that the compliance matrix reduces to the matrix

$$s = \begin{pmatrix} s_{11} & s_{12} & s_{13} & s_{14} & 0 & 0 \\ s_{12} & s_{11} & s_{13} & -s_{14} & 0 & 0 \\ s_{13} & s_{13} & s_{33} & 0 & 0 & 0 \\ s_{14} & -s_{14} & 0 & s_{44} & 0 & 0 \\ 0 & 0 & 0 & 0 & s_{44} & 2s_{14} \\ 0 & 0 & 0 & 0 & 2s_{14} & 2(s_{11}-s_{12}) \end{pmatrix} \quad (12b)$$

The c matrix is found in the same way to be

$$c = \begin{pmatrix} c_{11} & c_{12} & c_{13} & c_{14} & 0 & 0 \\ c_{12} & c_{11} & c_{13} & -c_{14} & 0 & 0 \\ c_{13} & c_{13} & c_{33} & 0 & 0 & 0 \\ c_{14} & -c_{14} & 0 & c_{44} & 0 & 0 \\ 0 & 0 & 0 & 0 & c_{44} & c_{14} \\ 0 & 0 & 0 & 0 & c_{14} & \frac{1}{2}(c_{11}-c_{12}) \end{pmatrix} \quad (13b)$$

It is necessary to understand clearly the meaning of each term in the s and c matrices. The term s_{11} is the ratio[2] of an extensional strain in the X direction to a tensile stress in the same direction. The terms s_{12} and s_{13} relate a tensile strain in the X direction with extensional stresses in the Y and Z directions, respectively. The ratio s_{12}/s_{11} and the ratio s_{13}/s_{11} are analogous to the quantity known as Poisson's ratio in isotropic solids. The algebraic signs of the various terms must, of course, be appropriate to the conventions used in defining the quantities.

The meaning of s_{14} is somewhat more difficult to visualize. A nonzero value of s_{14} means that a shear stress about the X axis results in an extensional strain in the X direction. Such effects do not occur in isotropic materials but are common in crystals. One may visualize the effect by considering the change of length of a helical spring as it is wound or unwound. The zero values of s_{15} and s_{16}, in quartz, mean that shear stresses about the Y- and Z-axes produce no extensional strain in the X-direction.

In the same manner, s_{21} relates an extensional strain in the Y direction to a tensile stress in the X direction. The numerical value of s_{21} must be equal to that of s_{12} in order to satisfy the law of conservation of energy. The value of s_{22}, which relates the extensional strain in the Y direction to the tensile stress in the Y direction, is equal to s_{11} in quartz. (It may be worth noting that when a subscript ij is a multiple of 11, the stress and strain are parallel. But it must be noted that a subscript in a matrix term should not be read as a number. For instance s_{45} should be read "s four five" and not "s forty-five.")

The remaining coefficients of the s matrix have similar interpretations. The terms in the c matrix are interpreted in the same way. The zero terms are especially worthy of some attention because they tell which stresses and strains are unrelated. Conversely, the nonzero value of $s_{56} = 2s_{14}$ is particularly troublesome, since it shows that a shear stress about the Z axis produces a shear strain about the Y axis. Examples such as this one are termed *mechanical coupling.* The importance of this type of coupling will be discussed in the section on "Activity Dips."

---

[2] The coefficient s_{ij} is not a ratio in the mathematical sense, since it is the quotient of a strain by a stress and is therefore not a dimensionless quantity.

Many interesting and important relationships between the coefficients can be observed. For example, $s_{11} = s_{22}$ but not s_{33}. $s_{44} = s_{55}$ but not s_{66}.

It can be seen in a semiquantitative way why $s_{66} = 2(s_{11} - s_{12})$. Consider a Z-cut section of quartz as shown in Fig. 3.6. The Z-axis is perpendicular to the page. The sides of the square $ABCD$ are parallel to the X- and Y-axes. If the quartz is subjected to an X_y strain, the square $ABCD$ is distorted into rhombus $A'B'C'D'$ and the square $MNOP$ is distorted into the rectangle $M'N'O'P'$. The strains are equivalent. The stresses involved in distorting $MNOP$ into a rectangle involve the coefficients s_{11} and s_{12}, which are independent of direction in the XY-plane. The number 2 arises from the fact that the longitudinal strain is referred to the distance MN, while the pure shear strain is measured by the angular rotation of line BC about point N and not about point B.

THERMAL STRESSES AND STRAINS

Strains in a solid such as quartz may be produced by internal as well as external forces. One method by which strain is induced is change of temperature.

A change of temperature produces not only changes of dimensions but also, in certain cases, changes of angle. The thermal stresses can be calculated by imagining the crystal to be clamped in such a way that neither its size nor shape can change. We might, for example, imagine it to be encased in a perfectly rigid casting. The stresses could

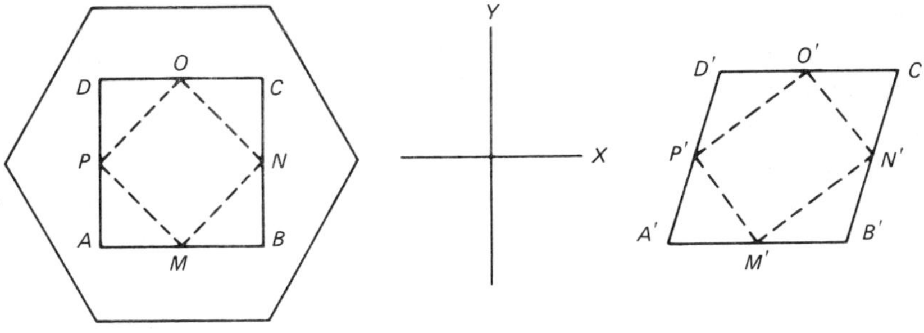

Fig. 3.6. The effect of X_y shear.

be calculated by first calculating the thermal strain and then the mechanical stress required to suppress that strain.

As a very simple one-dimensional example, suppose that the temperature of an *X*-cut plate is increased 1°. The thickness would increase by 14.3×10^{-6} cm for each centimeter of thickness. Now imagine that a force is applied sufficiently large to compress the blank back to its initial thickness. The elastic modulus c_{11} is 86.05×10^9 N/m^2. Hence a compressional stress of 86×10^9 N/m^2 $\times 14.3 \times 10^{-6} = 1.2 \times 10^6$ N/m^2 would be required to prevent thermal expansion resulting from a temperature increase of 1°. This is equivalent to a pressure of approximately 12 atm.

From considerations of this kind it is clear that extremely large stresses may result if different parts of the crystal are heated to different temperatures. Thermal gradients of only a few tenths of a degree per centimeter can produce stresses great enough to exceed the fracture point and cause the crystal to break. Another effect of stress, whether induced mechanically or thermally or by a combination of the two, is to produce electrical twinning in the quartz crystal.

A quartz crystal must never be heated or cooled rapidly. The larger the crystal, the less rapidly may the temperature be safely changed. The thermal conductivity of quartz is low and strongly anisotropic. Therefore, the time required for heat flow to equalize the temperature throughout the crystal is long. For example, a quartz crystal having a mass of 250 g (0.55 lb) should not be heated or cooled at a rate exceeding a few (say 5) degrees per minute.

Internal stresses and strains can also be produced in quartz by the application of an electric field. This is called the piezoelectric effect which is the subject of the next chapter.

Note that in Table 3.1 opposite signs are given to c_{14} and s_{14} by the two authorities. All older authorities, including Mason, Cady, and Vigoreux have followed Voigt in assigning c_{14} a positive sign.

A list of the equations required for computing the values of the effective elastic and piezoelectric coefficients for a rotation of θ about the *X*-axis has been given by Sykes[3]. These equations require that c_{14} be assigned a positive sign. For example, the equation for c'_{66},

---

[3]Raymond A. Heising, *Quartz Crystals for Electrical Circuits.* New York: Van Nostrand, 1946, pp. 247–248.

Table 3.1. The Elastic Coefficients of Alpha Quartz.

	Stiffness			Compliance	
	Bechmann*	Mason†		Bechmann*	Mason†
c_{11}	86.74	86.05×10^9 N/m²	s_{11}	12.77	12.79×10^{-12} m²/N
c_{12}	6.99	5.05	s_{12}	-1.79	-1.535
c_{13}	11.91	10.45	s_{13}	-1.22	-1.10
c_{14}	-17.91	18.25	s_{14}	+4.50	-4.46
c_{33}	107.2	107.1	s_{33}	9.60	9.56
c_{44}	57.94	58.65	s_{44}	20.04	19.78
c_{66}	39.88	40.5	s_{66}	29.12	28.65
	10^9 N/m² = 10^{10} dyn/cm²			10^{-12} m²/N = 10^{-13} cm²/dyn	

*R. Bechmann, *Phys. Rev.*, **110**, (5): 1060–1061 (June 1958).

†W. P. Mason, *Piezoelectric Crystals and their Applications to Ultrasonics*, New York: Van Nostrand, 1950, p. 84.

which is the effective stiffness coefficient for the AT- and BT-cut piezoids, is

$$c'_{66} = c_{44}\, s^2 - 2c_{14}\, sc + c_{66}\, c^2$$

where $c = \cos\theta$ and $s = \sin\theta$. In order to agree with the observed values, c_{14} must be assigned a positive sign.

However, computation of the rotation matrix for c'_{66} gives

$$c'_{66} = c_{44}\, s^2 + 2c_{14}\, sc + c_{66}\, c^2$$

which requires c_{14} to have a negative sign.

Later authorities including Beckmann et al. and Mindlin et al.[4] assign the minus sign to c_{14} and the plus sign to s_{14}. The origin of the discrepancy is not clear but in this book, following Bechmann, the value of c_{14} is taken to be negative.

[4]R. Bechmann, A. D. Ballato, and T. J. Lukaszek, *Proc. I.R.E.*, **50**:1816 (1962); R. D. Mindlin, and D. C. Gazis, *ibid.*

4
THE PIEZOELECTRIC EFFECT

INTRODUCTION

Piezoelectric crystals, particularly quartz, are used to control and manage the frequencies of the carrier signals used in electric communications. No other method of frequency control has been devised which competes with the vibrating quartz piezoid in accuracy, convenience, cost, and reliability. The quartz crystal unit provides coupling between the mechanical resonance frequency of the piezoid and the electric circuit, i.e., it is an electromechanical transducer. In doing so an extremely small amount of power is dissipated, with the result that the resonant frequency is very precisely determined. Finally, the resonant frequency is dependent on the dimensions of the piezoid and the mechanical and physical properties of quartz, which are highly stable.

THEORY OF PIEZOELECTRICITY

The coupling between the mechanical properties of the quartz and the electric circuit is through the piezoelectric effect. Although the piezoelectric effect was discovered in 1880 and the mathematical theory of the effect is completely established, little has been written about the origin of the effect. The latest books on the physics of solids seldom devote more than a paragraph to the subject and, if mentioned at all in textbooks on general physics, the treatment is usually superficial and/or confused with electrostriction.

To understand the origin of the piezoelectric effect, we first consider an electric dipole. A dipole is two charges of the same magnitude and opposite sign separated by a fixed distance. A dipole consisting of the charges $+q$ and $-q$ separated by a distance d is shown in Fig. 4.1. The dipole moment is defined as the product of the charge and the separation. The dipole moment of the dipole of Fig. 4.1 is therefore qd. The dipole moment is a vector quantity having the direction of the line joining the negative charge to the positive charge.

We will now compute the electric-field intensity at a point P at a distance r from the point midway between the charges along a line joining the two. We assume that $r \gg d$.

The electric field may be computed by application of Coulomb's law, but it is easier to compute first the electrostatic potential at point P and then to compute the field (intensity) by using $-E = \partial V/\partial r$. The potential due to a charge q at a point situated a distance r from the charge is very simply given by

$$V_p = \frac{4\pi\epsilon_o q}{r} = \frac{kq}{r} \qquad k = 4\pi\epsilon_o$$

The potential at a point due to several charges is the algebraic sum of the contributions of the several charges.

The potential of P, Fig. 4.1, is

$$V = k\left(\frac{+q}{r-a} + \frac{-q}{r+a}\right) = kq\,\frac{2a}{r^2-a^2}$$

and if a^2 is negligible compared with r,

$$V = \frac{2kqa}{r^2}$$

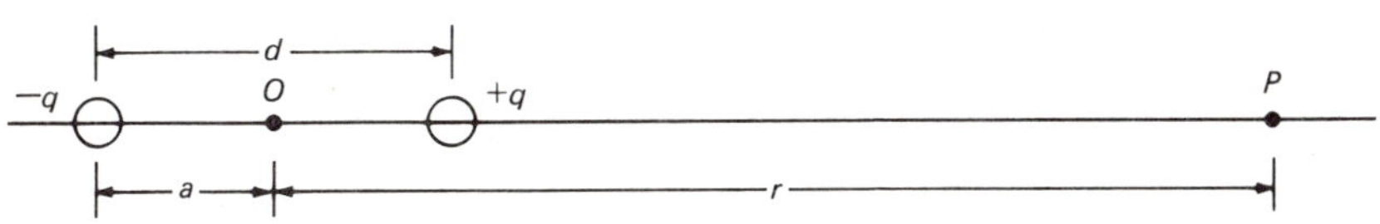

Fig. 4.1. An electric dipole.

The electric-field intensity is then

$$E = -\frac{\partial V}{\partial r} = \frac{4kqa}{r^3} = \frac{2kqd}{r^3} \tag{14}$$

Equation (14) shows that the electric-field intensity of a dipole, at a distant point along the axis of the dipole, is proportional to the dipole moment. Any change in the dipole must, therefore, cause a corresponding change in the electric field. This is the origin of the piezoelectric effect.

It is interesting and instructive to consider a slightly more complicated two-dimensional model. Suppose that six charges are located at the corners of a regular hexagon, as shown in Fig. 4.2. We take the X axis to be parallel to the line joining opposite charges and let a be the distance between the charges along this line.

We compute the potential V at a point P at a distance $OP = r$, where we assume $r \gg a$. The contribution to the potential due to charge q_1 is $kq/(r-a/2)$. The contributions from q_2 and q_3 are each given by $kq/(r+a/4)$ if $OP = r$ is so large compared with a that a line from q_2 to P can be considered practically parallel to OP.

The potential at P due to the three positive charges is therefore

$$V_+ = k\left(\frac{q}{(r-a/2)} + \frac{2q}{(r+a/4)}\right)$$

$$= +\frac{3kq}{r}$$

after neglecting a term $a^2/8$.

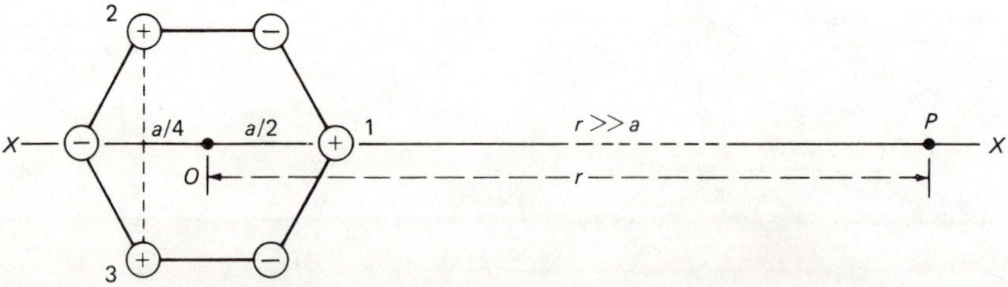

Fig. 4.2. Two-dimensional configuration of electric charges.

This is exactly the potential at P due to a charge of $+3q$ located at point O. Thus we have the important result that the potential, and the field intensity, at a distant point is the same for the three charges as for three coincident charges at the "center of gravity." Naturally the same result holds for the three negative charges, so that the potential and field at P are zero because the two contributions cancel.

Now suppose, however, that forces are applied along the Y axis causing the system of charges to be distorted, as shown in Fig. 4.3*a*. The center of gravity of the three positive charges moves to the left and that of the three negative charges moves to the right. If we assume that the strain causes the line from O to charge q_2 to be rotated counterclockwise by a small angle $d\theta$, it is easy to show that the X component of the displacement of q_2 is given by $(a/2)$ cos $30° \; d\theta$. The same is true of charge q_3. It is then easy to see that the center of gravity of the three positive charges has been displaced to the left by an amount $(\sqrt{3}\, a \, d\theta)/2$ and that of the three negative charges to the right by the same amount. Thus a dipole has been created having a dipole moment of

$$p = 3\sqrt{3}\, qa \, d\theta \qquad \text{coulomb meters in mks units}$$

If we imagine a crystal made up of molecules consisting of ions arranged like the charges of Fig. 4.3*a*; that the crystal contains N

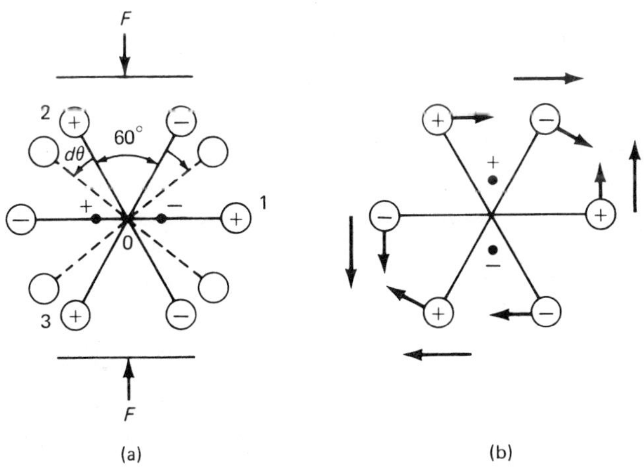

Fig. 4.3. The effects of (*a*) extensional and (*b*) shear strains.

such molecules per unit volume; and that each molecule is subjected to the same strain; then the dipole moment per unit volume is

$$P = Np = 3\sqrt{3}\ N\, qa\, d\theta$$

Although the model which we have been considering is hypothetical and certainly cannot agree in detail with a real crystal, it is of some interest to compute the piezoelectric coefficient e which it would have. The coefficient e is defined by the equation $P = e\,x$ [see Eq. (16)], where P is the polarization and x is the strain, which in this case is $d\theta$. Therefore, in our simple model,

$$e = 3\sqrt{3}\, N\, q\, a$$

If we take a to be 2.5 Å = 2.5×10^{-10} m, $N = 6 \times 10^{27}\ \mathrm{m}^{-3}$, both of which are typical values, and $q = 1.6 \times 10^{-19}$ C we obtain

$$e = 1.2\ \mathrm{C/m^2}$$

For purposes of comparison the value of e for various crystals are: quartz, $e_{11} = 0.173$, $e_{14} = 0.040$; Rochelle salt, $e_{25} = 0.15$, $e_{36} = 0.10$; tourmaline, $e_{15} = 0.25$, $e_{22} = 0.018$, $e_{31} = 0.103$, $e_{33} = 0.32$. Thus the value of e computed from even such a simple model differs by a factor of the order of 10 from the actual values for typical crystals.

The foregoing discussion is intended merely to give the reader a feel for the origin of the piezoelectric effect. It should not be expected that results derived from such a simple model should agree in detail or magnitude with real crystals.

It is possible to calculate the effect of a shear strain on the arrangement of charges shown in Fig. 4.3*b*. The pure shear, as we have seen in Fig. 3.5, is equivalent to two extensions at right angles. It can be seen from Fig. 4.3*b* that the effect of the shear strain is to cause the center of gravity of the three positive charges to move upward and that of the three negative charges to move downward, again producing a dipole. Calculation of the resulting piezoelectric coefficient is left as an exercise for the reader.

Before leaving the topic, we now show that the polarization (or dipole moment per unit volume) is equal to the piezoelectric displacement (or the surface charge per unit area). This important result can be seen in the following way. Imagine a rectangular parallelepiped of piezoelectric material. Let the surface area be A and the thickness be L. Let the sample contain n dipoles, each having a dipole moment $p = q\,d$ aligned in the direction of L.

The dipole moment per unit volume is obviously given by

$$P = \frac{nqd}{AL}$$

Now the charge per unit area on the surface A is equal to q times the number of charges which are exposed on the surface which is obviously $[n(d/L)]/A$. Therefore the charge per unit area is

$$\sigma = \frac{nqd}{AL}$$

which is the same as P.

We have been discussing the origin of the direct piezoelectric effect, i.e., the polarization resulting from a mechanical strain in a crystal. Within practical limits the polarization can be considered to be proportional to the strain and we may write

$$P = e\,x$$

where P is the polarization and x is the strain. However, as with Hooke's law in elasticity, the linear approximation is valid only for sufficiently small strain. In the design of quartz crystal units the linear approximation is adequate, although such is not necessarily the case with other materials.

THE CONVERSE PIEZOELECTRIC EFFECT

A piezoelectric crystal placed in an electric field may experience one or more kinds of mechanical strain. Again, within limits, the strain can be considered to be proportional to the applied field and we may write

$$x = d\,E$$

where x is the strain and E is the electric field. The constant d is called the *piezoelectric strain coefficient.*

The piezoelectric strains originate in the forces which charged particles experience in an electric field. Positive and negative ions tend to be pushed in opposite directions in an electric field and a dipole tends to rotate to align itself with the direction of the field. The resulting displacements give rise to the mechanical strains known as the converse piezoelectric effect.

Piezoelectric strains are usually quite small. For example, suppose that a potential difference of 100 V is applied to metallic electrodes deposited on the major surfaces of an X-cut plate of quartz having a thickness of 1.00 mm. The field is thus 1.00×10^5 V/m. The resulting strain is therefore $x = 2.27 \times 10^{-12}$ m/V $\times$ 1.00×10^5 V/m = 2.27×10^{-7}. This is the same strain which would be produced in the thickness of the plate by a temperature change of 0.016°C.

THE GENERALIZED PIEZOELECTRIC EQUATIONS

In the example just given, a field in the X direction produces an extensional strain in the X direction. The same electric field might (and often does) produce extensional strains in the Y and Z directions. Moreover, the same electric field may produce shear strains about one or more of the axes. All possible linear relations between the three components of the electric field and the six components of strain are included in the following set of equations.

$$\begin{aligned}
x_x &= d_{11}E_x + d_{21}E_y + d_{31}E_z \\
y_y &= d_{12}E_x + d_{22}E_y + d_{32}E_z \\
z_z &= d_{13}E_x + d_{23}E_y + d_{33}E_z \\
y_z &= d_{14}E_x + d_{24}E_y + d_{34}E_z \\
z_x &= d_{15}E_x + d_{25}E_y + d_{35}E_z \\
x_y &= d_{16}E_x + d_{26}E_y + d_{36}E_z
\end{aligned} \qquad (15a)$$

where $x_x, \ldots, x_y$ are the extensional and shear strains and E_x, E_y, and E_z are the components of the electric field. The d_{ij} are the piezoelectric strain coefficients. Equation (15) may be written in matrix form

$$x = d\,E \tag{15b}$$

The three components of polarization may be written in terms of the six components of strain in the following way.

$$\begin{aligned} P_x &= e_{11}x_x + e_{12}y_y + e_{13}z_z + e_{14}y_z + e_{15}z_x + e_{16}x_y \\ P_y &= e_{21}x_x + e_{22}y_y + e_{23}z_z + e_{24}y_z + e_{25}z_x + e_{26}x_y \\ P_z &= e_{31}x_x + e_{32}y_y + e_{33}z_z + e_{34}y_z + e_{35}z_x + e_{36}x_y \end{aligned} \tag{16a}$$

In matrix notation,

$$P = e\,x \tag{16b}$$

Equations (15) and (16) relate the polarization and the field to the strain. It is sometimes necessary to have the equations relating the polarization and the field to the stress. We already have

$$\begin{aligned} x &= d\,E && \text{from Eq. (15}b\text{)} \\ P &= e\,x && \text{from Eq. (16}a\text{)} \\ -x &= s\,X && \text{from Eq. (12}a\text{)} \\ -X &= c\,x && \text{from Eq. (13}a\text{)} \end{aligned}$$

Eliminating x between the second and third equation gives

$$-P = es\,X \tag{17}$$

And eliminating x between the first and fourth equation gives

$$-X = cd\,E \tag{18}$$

Equations (17) and (18) are the required equations, but we can show that $es = d$ and $cd = e$, so that we can write

$$-P = d\,X \quad \text{and} \quad -X = eE$$

The proof is a little complex depending on arguments from thermodynamics. We imagine a piezoelectric crystal to be placed in an electric field and at the same time subjected to a mechanical stress. The temperature is held constant. The electric field E produces polarization P and the stress X produces a strain x. Both represent energy stored in the crystal called the internal energy U. We now imagine the field to be changed by an infinitesimal amount dE and the stress by an infinitesimal amount dX. The change in the internal energy is[1]

$$dU = P\,dE - x\,dX \tag{19}$$

$$\left(\frac{\partial U}{\partial E}\right)_X = P \qquad \text{and} \qquad \left(\frac{\partial U}{\partial X}\right)_E = -x$$

$$\frac{\partial^2 U}{\partial E \partial X} = \frac{\partial p}{\partial X} \qquad \text{and} \qquad \frac{\partial^2 U}{\partial X \partial E} = -\frac{\partial x}{\partial E}$$

If the process is reversible, i.e., if energy is conserved, then Eq. (19) is an exact differential and the order of differentiation is immaterial; therefore,

$$\frac{\partial P}{\partial X} = -\frac{\partial x}{\partial E}$$

But from Eq. (17),

$$\frac{\partial P}{\partial X} = -es$$

and from Eq. (15*b*),

$$\frac{\partial x}{\partial E} = d$$

so that we have

$$d = es$$

By writing $dU = P\,dE - X\,dx$ and using the same procedure, one can show that $e = cd$. We may now write the piezoelectric stress equations as follows.

[1] The reader who is familiar with the subject of thermodynamics will recognize the analogy between Eq. (19) and the enthalpy equation for a reversible system.

$$\begin{aligned} -X_x &= e_{11}E_x + e_{21}E_y + e_{31}E_z \\ -Y_y &= e_{12}E_x + e_{22}E_y + e_{32}E_z \\ -Z_z &= e_{13}E_x + e_{23}E_y + e_{33}E_z \\ -Y_z &= e_{14}E_x + e_{24}E_y + e_{34}E_z \\ -Z_x &= e_{15}E_x + e_{25}E_y + e_{35}E_z \\ -X_y &= e_{16}E_x + e_{26}E_y + e_{36}E_z \end{aligned} \tag{20}$$

and

$$\begin{aligned} -P_x &= d_{11}X_x + d_{12}Y_y + d_{13}Z_z + d_{14}Y_z + d_{15}Z_x + d_{16}X_y \\ -P_y &= d_{21}X_x + d_{22}Y_y + d_{23}Z_z + d_{24}Y_z + d_{25}Z_x + d_{26}X_y \\ -P_z &= d_{31}X_x + d_{32}Y_y + d_{33}Z_z + d_{34}Y_z + d_{35}Z_x + d_{36}X_y \end{aligned} \tag{21}$$

To summarize, we have (written in matrix notation)

$$x = d\,E \tag{22}$$

$$-P = d\,X \tag{23}$$

$$-X = e\,E \tag{24}$$

$$P = e\,x \tag{25}$$

where Eq. (22) is shorthand for

$$\begin{pmatrix} x_x \\ y_y \\ z_z \\ y_z \\ z_x \\ x_y \end{pmatrix} = \begin{pmatrix} d_{11} & d_{21} & d_{31} \\ d_{12} & d_{22} & d_{32} \\ d_{13} & d_{23} & d_{33} \\ d_{14} & d_{24} & d_{34} \\ d_{15} & d_{25} & d_{35} \\ d_{16} & d_{26} & d_{36} \end{pmatrix} \begin{pmatrix} E_x \\ E_y \\ E_z \end{pmatrix} \tag{22a}$$

and Eq. (23) means

$$-\begin{pmatrix} P_x \\ P_y \\ P_z \end{pmatrix} = \begin{pmatrix} d_{11} & d_{12} & d_{13} & d_{14} & d_{15} & d_{16} \\ d_{21} & d_{22} & d_{23} & d_{24} & d_{25} & d_{26} \\ d_{31} & d_{32} & d_{33} & d_{34} & d_{35} & d_{36} \end{pmatrix} \begin{pmatrix} X_x \\ Y_y \\ Z_z \\ Y_z \\ Z_x \\ X_y \end{pmatrix} \tag{23a}$$

Note that the rows and columns are interchanged in the d matrices of Eqs. (22*a*) and (23*a*). This is necessary in order to be able to perform the required multiplications. Not being square matrices, neither the *d* nor *e* matrix has a reciprocal matrix, so it is not possible to "solve" for the variables as in the case of the square matrix.

THE PIEZOELECTRIC MATRIX FOR QUARTZ

Although the piezoelectric matrix contains 18 terms in the general case, symmetry conditions usually require many of the terms to be zero and establish relations between others. To determine the relationships between the terms in the piezoelectric strain matrix for quartz it is sufficient to perform a rotation of 120° about the Z axis followed by a rotation of 180° about the X axis.

The strain matrix is rotated by use of the appropriate matrix[2] to obtain

$$x' = \alpha_3 x$$

where α_3 is the 6 X 6 matrix for rotation about the Z axis.

The electric field, being a vector, is rotated by the 3 X 3 matrix of Eq. (10*c*). We then have

$$E' = c E \qquad \text{or} \qquad E = c^{-1} E'$$

Substituting in the piezoelectric strain matrix from Eq. (22) we have

$$x' = \alpha_3 x = \alpha_3 d E = \alpha_3 d c^{-1} E' = d'E' \qquad \text{where } d' = \alpha_3 d c^{-1}$$

Repeating the same process for a rotation of 180° about the X axis we have

$$\begin{aligned} x'' &= \alpha_1 x' \\ E'' &= a E' \qquad \text{and} \qquad E' = a^{-1} E'' \\ x' &= d' E' \end{aligned}$$

from which

$$x'' = \alpha_1 x' = \alpha_1 d'E' = \alpha_1 d'a^{-1} E'' = \alpha_1 \alpha_3 d c^{-1} a^{-1} E''$$

$$\text{or} \qquad x'' = d''E'' \qquad \text{where} \qquad d'' = \alpha_1 \alpha_3 d c^{-1} a^{-1}$$

---

[2] See Walter L. Bond, "The Mathematics of the Physical Properties of Crystals," *B.S.T.J.*, 22: 1-72 (January 1943). This article includes tables of the matrices required to perform the rotations required in work in piezoelectricity.

Now the double rotation which is compatible with the symmetry of quartz must leave the piezoelectric matrix unchanged. Therefore, each term of d'' must be equal to the corresponding term of the matrix d. This leads to 18 simultaneous equations, the solution of which requires that 13 of the d_{ij} be zero and that only 2 of the remaining coefficients be independent.

In this way the d matrix for quartz is found to be

$$d = \begin{pmatrix} d_{11} & 0 & 0 \\ -d_{11} & 0 & 0 \\ 0 & 0 & 0 \\ d_{14} & 0 & 0 \\ 0 & -d_{14} & 0 \\ 0 & -2d_{11} & 0 \end{pmatrix} \tag{26}$$

The d matrix for quartz thus contains five nonzero terms but only two of them are independent. The five nonzero terms tell us the nature of the strains which can be produced by the three components of the electric field. The X-component of the electric field produces three types of strain; an extensional strain in the X-direction, a (negative) extensional strain in the Y-direction, and a shear strain about the X-axis, i.e., in the YZ-plane. The Y-component of the field produces a shear strain about the Y- axis and a shear strain about the Z-axis. The latter is used in the Y-cut family of piezoids which includes the AT- and BT-cuts. The former is the basis of the strain utilized in the CT- and DT-cuts. The Z-component of the field produces no piezoelectric strain.

The e matrix can be calculated from the d matrix by using the equation $e = cd$ (see page 58). Using Eqs. (13b) and (26) we have

$$\begin{pmatrix} c_{11} & c_{12} & c_{13} & c_{14} & 0 & 0 \\ c_{12} & c_{11} & c_{13} & -c_{14} & 0 & 0 \\ c_{13} & c_{13} & c_{33} & 0 & 0 & 0 \\ c_{14} & -c_{14} & 0 & c_{44} & 0 & 0 \\ 0 & 0 & 0 & 0 & c_{44} & c_{14} \\ 0 & 0 & 0 & 0 & c_{14} & \frac{c_{11}-c_{12}}{2} \end{pmatrix} \begin{pmatrix} d_{11} & 0 & 0 \\ -d_{11} & 0 & 0 \\ 0 & 0 & 0 \\ d_{14} & 0 & 0 \\ 0 & -d_{11} & 0 \\ 0 & -2d_{11} & 0 \end{pmatrix} = \begin{pmatrix} e_{11} & 0 & 0 \\ -e_{11} & 0 & 0 \\ 0 & 0 & 0 \\ e_{14} & 0 & 0 \\ 0 & -e_{14} & 0 \\ 0 & -e_{11} & 0 \end{pmatrix}$$

By carrying out the indicated operations it is seen that

$$e_{11} = (c_{11} - c_{12})d_{11} + c_{14}\, d_{14} \quad \text{and} \quad e_{14} = 2c_{14}\, d_{11} + c_{44}\, d_{14}$$

Hence, if we measure d_{11} and d_{14} and know the elastic constants c_{11}, c_{12}, c_{14}, and c_{44}, we can calculate all the terms of the e matrix.

We show in Chap. 6 that the characteristics of AT- and BT-cut resonators depend upon the value of e'_{26} which is obtained by rotating the e matrix about the X axis by the angle θ. The value is found to be

$$e'_{26} = (e_{14}\, sc - e_{11}\, c)c$$

where c = cos θ and s = sin θ.

Note: In Chaps. 6 and 7 the symbol e for the piezoelectric strain coefficient is replaced with ϵ to avoid confusion with the symbol e, which designates thickness.

Table 4.1. The Piezoelectric Coefficients of Alpha Quartz.

$d_{11} = +2.27 \times 10^{-12}$ m/V	$e_{11} = +0.173$ C/m^2
$d_{14} = -0.67$	$e_{14} = +0.040$

5
MECHANICAL VIBRATIONS

INTRODUCTION

In this chapter we discuss the ideal case of free vibrations of solid bodies. Although the quartz crystal unit is not a freely vibrating system, many of the pertinent features can be understood by considering the simpler case of free vibrations. The effects of frictional and driving forces will be considered later.

FUNCTION OF THE PIEZOID IN OSCILLATOR CIRCUITS

The quartz crystal unit serves the same purpose in an oscillator circuit as does the balance wheel and hair spring in a watch. In both cases the function is to provide isochronous (same time) vibrations, i.e., to mark off equal intervals of time.

The time required for the balance wheel of a watch to complete one vibration is given by

$$T = 2\pi \sqrt{\frac{I}{h}}$$

where I is the moment of inertia of the balance wheel and h is the stiffness coefficient of the hair spring. The equation above applies only to a freely vibrating system, i.e., one which is frictionless and not being driven. Of course, no vibrating system is frictionless, so if sustained oscillations are required, energy must be supplied to replace that lost in friction and other damping effects. The energy to replace that lost through damping is supplied by the main spring through the escapement.

The function of the quartz crystal unit in an oscillator circuit is likewise to mark off equal intervals of time or, in other words, to stabilize the frequency of oscillation. As with all mechanical vibrating systems, some energy is dissipated in the vibration of the crystal and this energy must be replaced if oscillations are to be sustained. The energy is supplied to the vibrating crystal through the piezoelectric effect, which is therefore analogous to the escapement of the watch.

Just as the period or frequency of the balance wheel of the watch is affected by friction and driving forces, so is the frequency of the quartz resonator similarly affected. The effect is small in both cases but especially so in a well-made quartz resonator where the fraction of the energy dissipated per cycle may be as low as one millionth.

Isochronous mechanical vibrations could presumably be excited in almost any solid, elastic material. However, the piezoelectric effect provides an extremely simple and convenient method of exciting vibrations in a material such as quartz in which a linear mechanical strain is created by an electric field. The properties of quartz which make it the most satisfactory material for use as a frequency-stabilizing element are its hardness, low internal dissipation, durability, uniformity, freedom from flaws, and not least, the combination of characteristics which makes possible the design of piezoids in which the resonant frequencies are substantially independent of temperature. Other crystals are known which satisfy some of these requirements but none, except quartz, satisfies all of them.

ONE-DIMENSIONAL VIBRATIONS

Any solid body may be set into vibration in many different ways, that is, it has many resonant frequencies. Various types of vibrations are used in piezoelectric resonators. In every case a natural resonant frequency of the mechanical system is used to control the frequency of the associated electric system. The simplest type of ideal vibrating system is one in which the vibrating body has only one dimension. Naturally no such device exists, but several systems may be treated as one-dimensional, at least to a high degree of approximation. Whenever the controlling dimension is very large or very small compared with the other dimensions, the vibration can be considered to be approximately one-dimensional.

Probably the simplest example of a one-dimensional system is provided by a thin, flexible string stretched between two fixed points. Another example is a long, thin elastic rod vibrating in such a way that the displacement is parallel to the length of the rod. A long, slender organ pipe is a fairly good approximation to a one-dimensional system. A plate in which the length and width are very large compared with the thickness can be treated as a one-dimensional system for waves traveling in the thickness direction.

The differential equation describing the free vibration of a one-dimensional system is

$$\frac{\partial^2 \psi}{\partial t^2} = \nu^2 \frac{\partial^2 \psi}{\partial x^2} \tag{27}$$

where

ψ = displacement of a point in the vibrating body relative to its equilibrium position,
ν = velocity with which a disturbance corresponding to the displacement ψ is propagated in the body,
x and t = space and time coordinates, respectively.

Equation (27) is applicable to one-dimensional systems in which the displacement ψ is either parallel or perpendicular to the direction of wave propagation.

To find a solution of Eq. (27) we assume that a solution can be found having the form

$$\psi(x, t) = [T(t)]\,[X(x)] \tag{28}$$

where T is a function only of time and X is a function of the space coordinate x. We take the partial derivatives of $\psi(x, t)$ with respect to t and x and substitute these into Eq. (27). After dividing both sides by XT, we obtain

$$\frac{1}{T}\frac{d^2 T}{dt^2} = \frac{\nu^2}{X}\frac{d^2 X}{dx^2} \tag{29}$$

The expression on the left side of Eq. (29) is a function only of t and that on the right only of x. It is necessary, therefore, that each be equal to the same constant which we will take as $-\omega^2$.

We now have

$$\frac{1}{T}\frac{d^2T}{dt^2} = -\omega^2 \quad \text{and} \quad \frac{v^2}{X}\frac{d^2X}{dx^2} = -\omega^2$$

The solution of the equation in t is

$$T = A \sin(\omega t) + B \cos(\omega t)$$

and that of the equation in x is

$$X = D \sin\left(\frac{\omega x}{v}\right) + E \cos\left(\frac{\omega x}{v}\right)$$

The product of these two equations is a solution of Eq. (27). The values of the arbitrary constants, A, B, C, D, and ω are determined by the mechanical constraints imposed upon the vibrating system. Such constraints are called *boundary conditions.*

Before the solution can be carried further, it is necessary to define the nature of the vibrating system. Let us suppose that we have a thin string of length L rigidly attached to points at $x = 0$ and $x = L$. We shall further assume that $\psi = 0$ when $t = 0$, which is equivalent to deciding when we are to start measuring time. Mathematically, the boundary conditions can be expressed as

1. $\psi = 0$ for $x = 0$
2. $\psi = 0$ for $x = L$
3. $\psi = 0$ for $t = 0$

Condition 1 requires that $E = 0$. Condition 2 requires that

$$\frac{\omega L}{v} = n\pi \qquad n = 1, 2, 3, \ldots$$

and condition 3 requires that $B = 0$.

A solution of Eq. (27) consistent with the three boundary conditions is therefore

$$\psi_n = A_n \sin\left(\frac{n\pi v}{L} t\right) \sin\left(\frac{n\pi}{L} x\right)$$

For our purpose it is significant to note that the string may vibrate in many different modes depending on the value of n which must be an integer. The frequency of the vibration in each of these modes is found by setting

$$\omega_n = \frac{n\pi v}{L} = 2\pi f_n \qquad \text{or} \qquad f_n = n\frac{v}{2L} \tag{30}$$

The quantity $f_1 = (v/2L)$ is called the fundamental frequency or the first harmonic. The frequency $f_2 = 2f_1$ is called the second harmonic, etc. Thus f_n, the frequency of the nth harmonic, is exactly n times the frequency of the fundamental and the "eigenfrequencies" or resonant frequencies of the simple one-dimensional system are related as the integers 1, 2, 3, ... Not only may the string vibrate at any one of these frequencies, it may also vibrate at all of them simultaneously. The characteristic tones of string musical instruments are the result of simultaneous vibrations at a number of different frequencies.

The displacement of the string vibrating in modes corresponding to $n = 1$, 2, and 3 is illustrated in Fig. 5.1. The relative amplitudes and phases of the various modes depend on the way in which the string is set into motion but the frequencies are determined by Eq. (30).

The modes of motion of the string are determined by the conditions that the ends of the string must be nodal points (points of zero displacement) and that the length of the string be an integral number

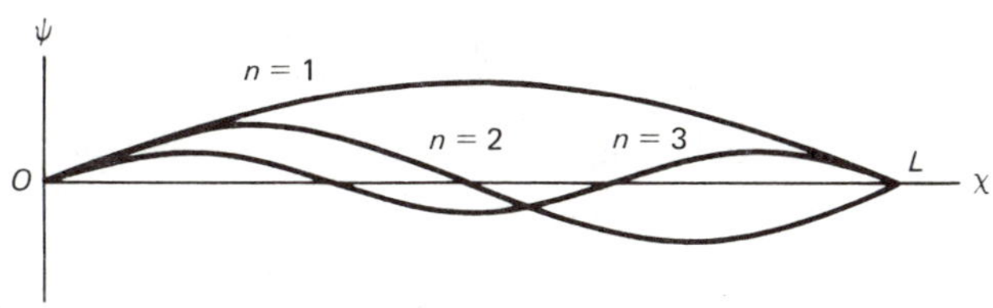

Fig. 5.1. Modes of vibration of a stretched string.

of half-wavelengths of the wave which travels from one end of the string to the other. These conditions, of course, serve to determine the resonant frequencies. The most complicated vibration which a string may execute is made up of a superposition of a set of modes. The amplitudes of the various modes as well as their phase relationships may be determined by writing the Fourier series for the shape of the string at the time it is set into motion.[1]

THE VIBRATING BAR

The solution of the problem of the longitudinal vibration of a long, thin bar is carried out in the same way done for the string. However, vibrating bars of quartz are usually supported at the midpoint rather than at the ends. The displacement is parallel to the length of the bar instead of at right angles as in the case of the string. The differential equation of motion is the same [Eq. (27)] and the solution is carried out in the same way except that the boundary conditions are different. Since the bar is clamped at its midpoint, that point must be a nodal point and the ends of the bar, being unconstrained, are antinodes. With these boundary conditions, the solution requires that

$$f_n = n\frac{v}{2L} \qquad n = 1, 3, 5, \ldots$$

and the resonant frequencies are related as the odd integers.

It is somewhat difficult to visualize the motion of the longitudinally vibrating bar. One way to do so is to imagine a series of dots to be placed on the surface of the bar before it is set into vibration. When the bar vibrates the dots appear to be stretched into lines by the phenomenon of persistence of vision. The lines would be longest at the antinodes and shortest near the nodal points. Figure 5.2 illustrates how the dots would appear when the bar is vibrating in modes $n = 1$, 3, and 5.

The length of the bar is $\lambda/2$, $3\lambda/2$, $5\lambda/2$, etc. No modes corresponding to $n = 2, 4, 6, \ldots$ exist.

---

[1] See R. L. Churchill, *Fourier Series and Boundary Value Problems.* McGraw-Hill, New York, 1969.

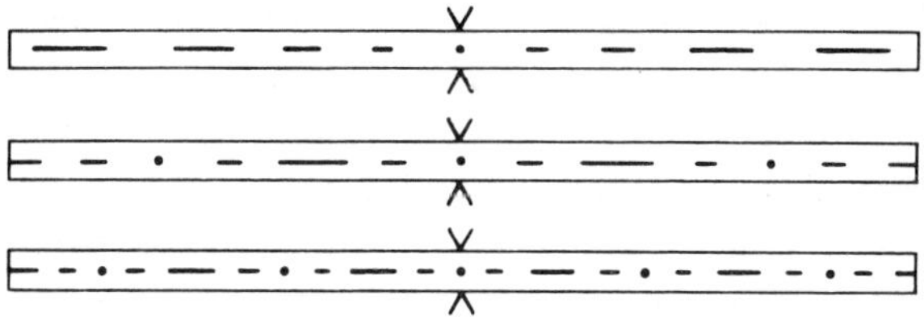

Fig. 5.2. Modes of motion of a vibrating bar clamped at its midpoint. The dots are nodal points.

Precise measurements of the resonant frequencies of actual vibrating systems reveal that they are not exactly integral multiples of the fundamental frequency, because actual vibrating systems are not perfectly one-dimensional. Every string has a finite thickness and every rod or bar has a finite width and thickness. If the length is very large compared with the lateral dimensions, the resonant frequencies closely approach the harmonic values.

TWO-DIMENSIONAL VIBRATING SYSTEMS

The CT- and DT-cut family of resonators consist of either a square or circular plate of quartz vibrating in a face-shear mode. The centerpoint of the plate is a nodal point and the plate is supported by attaching lead wires at that point. Such resonators closely approximate two-dimensional vibrating systems. It is instructive to investigate the vibration of such systems as a basis for understanding these devices and to illustrate further the principles of vibrating solids.

The differential equation of motion of a two-dimensional system, in rectangular coordinates, is

$$\frac{\partial^2 \psi}{\partial t^2} = v^2 \left(\frac{\partial^2 \psi}{\partial x^2} + \frac{\partial^2 \psi}{\partial y^2} \right) \tag{31}$$

where ψ is the displacement, either in the plane of the plate or perpendicular to it. The wave velocity v is the velocity of the wave traveling in the plane of the plate. The velocity v depends on the density and the applicable elastic constants. The quantities x, y, and t are the usual space and time coordinates. Equation (31) may be solved in exactly the same way that Eq. (27) was solved for the one-dimensional case.

If we assume that the plate is a rectangle with dimensions a and b and that the center of the plate is constrained to be a nodal point of displacement, then the solution of Eq. (31) consistent with the boundary conditions above is

$$\psi_{nm} = A_{nm}\left(\sin \frac{n\pi x}{a}\right) \sin\left(\frac{m\pi y}{b}\right) \sin\left(\omega_{nm} t\right)$$

where

$$\omega_{nm} = \pi v \sqrt{\frac{n^2}{a^2} + \frac{m^2}{b^2}} \qquad n, m = 1, 3, 5, \ldots \qquad (32)$$

or

$$f_{nm} = \frac{v}{2} \sqrt{\frac{n^2}{a^2} + \frac{m^2}{b^2}}$$

The relationship between the frequencies of the modes of the two-dimensional system is not as simple as in the one-dimensional case. The mode of lowest frequency is the mode in which $n = 1$ and $m = 1$. This is the fundamental frequency. The shape assumed by the plate when vibrating in the 11 (read one one) mode is shown in Fig. 5.3*b*. Because of the boundary conditions, the length and width of the plate must be an odd number of half-wavelengths of the wave propagated along the same direction. Hence the modal index 11 indicates that the length and width of the plate are each one half-wavelength.

The mode next higher in frequency depends on the length and width of the plate. The frequencies can be readily computed using

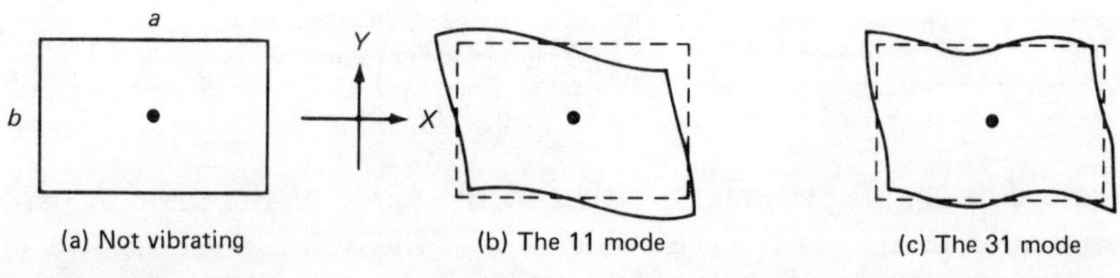

Fig. 5.3. Modes of motion of a thin plate vibrating in face shear. (*a*) not vibrating; (*b*) the 11 mode; (*c*) the 31 mode.

the formula derived above. For example, if $a = 3$ and $b = 2$, the relative frequencies of the first few modes are shown in Table 5.1.

From Table 5.1 it can be seen that some of the overtones are harmonically related to the fundamental mode but other modes are not. In the example given, the first overtone mode is the 31 mode, but its frequency is not an integral multiple of the frequency of the 11 mode. However, the frequency of the 33 mode is exactly three times that of the 11 mode.

Many quartz crystal units are made from square blanks vibrating in the face-shear mode. It is readily seen, from Eq. (32), that the frequency of the *nm* mode is given by

$$f_{nm} = \frac{v}{2a}\sqrt{n^2 + m^2} \qquad m, n = 1, 3, 5, \ldots \tag{33}$$

Hence the frequencies of the *nm* and *mn* modes are equal. In such cases the modes are said to be *degenerate*.

So far we have been discussing the vibration of isotropic plates. In crystals, however, the wave velocity is likely to differ with direction in the crystal. If we let the wave velocity in the Y direction be b times that in the X direction which is taken as v, then we have for a square anisotropic plate

$$f_{nm} = \frac{v}{2a}\sqrt{n^2 + b^2 m^2} \quad m, n = 1, 3, 5, \ldots \tag{34}$$

Piezoids, made from circular plates of quartz, can also be excited in face-shear modes. Since the solution of the equation of motion in polar coordinates involves Bessel's functions and is rather complicated,

Table 5.1. Relative Frequencies of the Overtone Modes of a Plate Vibrating in Face Shear.*

Mode	Relative Frequency
11	0.601
31	1.118
13	1.536
33	1.803
55	3.005

*Length =3, width = 2. To obtain actual frequencies, multiply by $v/2$.

we omit the details. The results show that a circular plate may also vibrate in many different modes, the frequencies of which are not harmonically related. The nodal lines are circular and radial with a node at the central point where the blank is supported.

The behavior of two-dimensional systems differs from that of one-dimensional systems in two respects. The number of modes is much larger in two-dimensional systems and the frequencies of the modes are not harmonically related as in the one-dimensional system. The inharmonic modes are responsible for the characteristic musical tones of bells and cymbals. They are also responsible for a number of problems in the design of quartz crystal resonators. These problems are discussed later.

Either an X-cut or a Y-cut plate of quartz can be excited into vibration in a face-shear mode by imposing an electric field of the appropriate frequency in the thickness direction of the plate. The X-cut plate is excited through the coefficient d_{14} of Eq. (26), which relates the field in the X-direction to a shear strain about the X-axis. The Y-cut plate is excited through the coefficient $d_{25} = -d_{14}$, which relates a field in the Y-direction to a shear strain about the Y-axis. Neither of these piezoids is particulary useful, because the frequencies are not temperature-independent. By rotating the plane of the Y-cut about the X-axis, two positions are found at which the frequency of the face-shear vibration is substantially independent of temperature. These are known as the CT- and DT-cuts. Quartz crystal units employing these cuts are extensively used in the frequency range between 100 and 1000 kHz.

THREE-DIMENSIONAL VIBRATING SYSTEMS (RECTANGULAR PLATES)

Most quartz resonators are made in the form of thin plates. Circular plates have certain advantages, including ease of production, and are preferred for most applications. The theory, however, is simpler in the case of rectangular plates, so we will use it to illustrate the important features of the three-dimensional vibrating system.

X-cut plates of quartz can be set into extensional vibration in the thickness direction through coefficient d_{11} of Eq. (26). Such plates are sometimes used as ultrasonic transducers but the temperature coefficient of frequency which is about $-20 \times 10^{-6}/°C$ makes them undesirable for use as frequency controlling elements.

Y-cut plates of quartz may be excited into a thickness-shear mode of vibration through coefficient $d_{26} = -2d_{14}$ [Eq. (26)]. The temperature coefficient of frequency of the Y-cut plate is about $+100 \times 10^{-6}/°C$ but by rotating the plane of the blank about the X-axis, two positions are found at which the frequency is nearly independent of temperature. The resulting piezoids are called the AT- and BT-cuts.

The mode of motion of the X-cut plate is equivalent to a longitudinal wave traveling in the thickness direction of the plate. The displacement in the Y-, AT-, and BT-cut plates is equivalent to a transverse wave traveling in the thickness direction of the plate. The equation of motion in both cases is

$$\frac{\partial^2 \psi}{\partial t^2} = v^2 \left[\frac{\partial^2 \psi}{\partial x^2} + \frac{\partial^2 \psi}{\partial y^2} + \frac{\partial^2 \psi}{\partial z^2}\right] \tag{35}$$

where the displacement ψ is taken to be parallel to the direction of wave propagation in the X-cut and perpendicular to the direction of propagation in the case of the Y-cut.

The solution of Eq. (35) is carried out by assuming that a solution in the form

$$\psi(x, y, z, t) = [X(x)]\ [Y(y)]\ [Z(z)]\ [T(t)]$$

can be found.

If we let the length l be in the X direction; the width w be in the Z direction; the thickness e in the Y direction and assume that the plate is not clamped, then the midpoint of the plate must be a node and the boundary conditions are

$$\psi = 0 \quad \text{at } x = 0$$
$$\psi = 0 \quad \text{at } y = 0$$
$$\psi = 0 \quad \text{at } z = 0$$

The solution of Eq. (35) satisfying these boundary conditions is

$$\psi_{nmp} = A_{nmp} \sin\left(\frac{n\pi y}{e}\right) \sin\left(\frac{m\pi x}{l}\right) \sin\left(\frac{p\pi z}{w}\right) \sin\left(\omega_{nmp} t\right) \tag{36}$$

where

$$\omega_{nmp} = \pi v \sqrt{\frac{n^2}{e^2} + \frac{m^2}{l^2} + \frac{p^2}{w^2}} \qquad n, m, p = 1, 3, 5, \ldots \tag{37}$$

The modes of motion thus consist of standing waves in the X, Y, and Z directions with the thickness, length, and width each being equal to an odd number of half-wave lengths. The corresponding frequencies given by Eq. (37) are, in general, not harmonically related. If l and w are both very large compared with the thickness e, then Eq. (37) can be written as

$$\omega_n = \frac{n\pi v}{e} \qquad \text{or} \qquad f_n = \frac{n\,v}{2e} \tag{38}$$

The wave-propagation velocity v is given by $v = \sqrt{c_{ij}/\rho}$, where $c_{ij} = c_{11}$ for the X-cut and c_{66} for the Y-cut. ρ is the density.

The resonant frequency of an infinitely large X-cut plate is therefore given by

$$f_n = \frac{n}{2e}\sqrt{\frac{c_{11}}{\rho}} = n\,\frac{2849}{e} \qquad (X\text{-cut}) \tag{39}$$[1]

and for an infinitely large Y-cut plate

$$f_n = \frac{n}{2e}\sqrt{\frac{c_{66}}{\rho}} = n\,\frac{1954}{e} \qquad (Y\text{-cut}) \tag{40}$$[1]

The overtone frequencies of an infinite plate would be odd-integral multiples of the fundamental frequency, but in plates of finite size, this is not the case. Furthermore, plates of finite size display many more modes of vibration, the frequencies of which depend upon the length and width of the plate and on the values of m and p. The mode of lowest frequency is the 111 mode. If, as is usually the case,

[1] In Eqs. (39) and (40), e is in meters and f is in hertz.

l and *w* are larger than *e*, then a series of modes appear, all of which have frequencies higher than the 111 mode. These modes have indices 131, 113, 133, 151, 153, 155, etc.

The inharmonic modes are the *spurs* (spurious modes) which are so troublesome in the design and fabrication of piezoids for use in high-frequency filters. Like the harmonic modes they are thickness-shear modes, controlled by the same elastic coefficients and having the same temperature coefficient of frequency. Figure 5.4 gives a qualitative picture of the distribution and relative amplitudes of the various modes of vibration of an AT-cut plate. In a later section we consider the steps which may be taken in the design and fabrication to minimize the effect of the overtone modes.

The inharmonic modes are not so important in the design of piezoids for oscillator applications. The harmonic modes are much easier to excite and the oscillator usually selects the strongest mode. There exist, however, many other modes of vibration which have different temperature coefficients of frequency. Occasionally, as the temperature of the piezoid is changed, the frequency of one of these modes may coincide with that of the oscillator frequency. The resulting extra dissipation of energy reduces the amplitude of oscillation (activity) and the result is called an *activity dip*. Sometimes the oscillator "jumps" to one of the inharmonic modes with a "skip" in frequency.

In a later section we discuss the problem of coupled modes and the steps which may be taken in the design of piezoids to minimize their effects.

In older designs for quartz crystal units, rectangular AT- and BT-cut plates were often placed between metal plates which served to clamp the plate around its edges or at its corners. These changes in the boundary conditions are such that the modal indices *m* and *p* may be even as well as odd integers and consequently modes such as

Fig. 5.4. Overtone spectrum of an AT-cut plate (qualitative).

112, 121, 123, etc., are permitted. The result is to increase greatly the number of inharmonic overtone modes.

THREE-DIMENSIONAL VIBRATING SYSTEMS (CIRCULAR PLATE)

For reasons which will be discussed later, most high-frequency quartz resonators are now made using circular plates. Although the general features of three-dimensional vibrating systems are similar for rectangular and circular plates, it is interesting and instructive to study them in some detail.

The complete solution of the equation of motion of a three-dimensional plate cut from an anisotropic crystal is extremely complex. The greatest mathematical difficulties arise from the uncertain nature of the boundary conditions. By neglecting the effects of anisotropy and idealizing the boundary conditions, a soluble problem is defined, and even though the solution yields numerical results which are only fair approximations, yet certain useful and important results are obtained.

Circular AT-cut plates are usually contoured to give one or both of the major surfaces a spherical shape. Sometimes a flat circular region is left in the center of the plate, as shown in Fig. 5.5*d*. Electrodes smaller than the quartz plate are deposited on the major surfaces concentric with the plate. Under these conditions the vibrating region is confined to a circular region which may be smaller, equal to, or larger than the area of the electrodes. The annular ring of quartz surrounding the vibrating area serves to confine the energy of vibration, giving rise to the "energy-trapping" concept which is treated in Chap. 10.

The problem is solved by writing the wave equation in cylindrical coordinates, separating the variables in the usual way, and applying

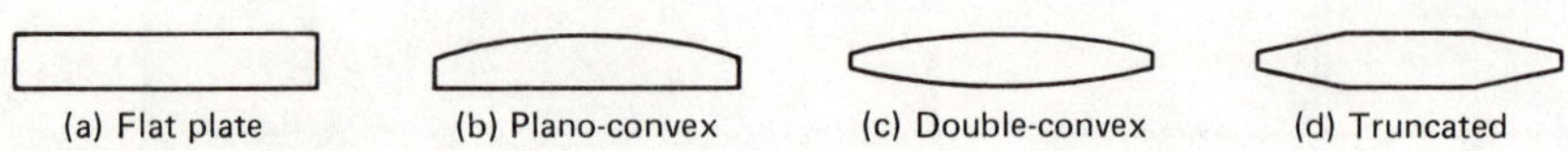

Fig. 5.5. Contoured quartz plates. (*a*) Flat plate; (*b*) plano-convex; (*c*) double convex; (*d*) truncated.

the boundary conditions described above. In this way the eigenfrequencies of the various normal modes are found to be

$$\omega_{nmk} = v \sqrt{\frac{n^2 \pi^2}{e^2} + \frac{\chi^2_{mk}}{a^2}} \qquad \begin{aligned} n &= 1, 3, 5, \ldots \\ m &= 0, 1, 2, 3, \ldots \\ k &= 1, 2, 3, \ldots \end{aligned} \tag{41}$$

and

$$f_{nmk} = \frac{\omega_{nmk}}{2\pi} \doteq \frac{nv}{2e} + \frac{v e \chi^2_{mk}}{4n(\pi a)^2} \tag{42}$$

In the equations above χ_{mk} is the kth root of Bessel's function of order m, values of which are found in tables of mathematical data.

The mode of the lowest frequency is the one for which $n = 1, m = 0$, and $k = 1$. If the radius of the plate a is very large compared with the thickness e, the frequency of the 101 mode is approximately $v/2e$, which is the same as that of the infinitely large rectangular plate. Obviously, if the boundary is infinitely far away, it does not matter whether it is circular or rectangular. Likewise, the frequencies of the 301, 501, etc., modes are approximately 3, 5, . . .times that of the fundamental mode, so we have

$$f_n = \frac{nv}{2e} = n \frac{K}{e} \qquad \text{where} \qquad K = \frac{v}{2} = \frac{1}{2} \sqrt{\frac{c_{ij}}{\rho}} \tag{43}$$

From Eq. (42) the frequencies of the inharmonic modes differ from that of the fundamental mode by terms which are proportional to the squares of the roots of Bessel's functions. The differences are also inversely proportional to the square of the radius of the vibrating area a. Therefore, if a is made smaller by contouring and/or depositing smaller electrodes, the frequency separation between the fundamental and inharmonic overtones is increased and this fact is utilized extensively in the design of crystal units for high-frequency filter applications.

It should not be expected that the frequencies of a given plate would be given exactly by Eq. (42) because of the approximations made in the definition of the problem. It is not surprising, however, that a large number of inharmonic modes may be observed. It is easy to measure the frequencies of the various modes by using the crystal unit as an element in a passive network and exciting it by a suitable oscillator, as shown in Fig. 5.6.

It is also possible to "observe" the displacement of the surface of the quartz plate by permitting some fine dust to fall on the surface of the quartz while it is excited piezoelectrically. The dust is driven from the regions of maximum motion to collect along the nodal lines. The resulting "dust patterns" clearly show the complicated nature of the vibration and the differences between the various modes of motion.

Figure 5.7 shows the actual overtone spectrum of a 3.1-MHz circular AT-cut plate having a diameter of 1.397 cm. The surfaces of the plate were lapped to have a radius of curvature of 25 cm (using a 2.0 diopter optical lap). Although the frequencies of the modes are not given precisely by Eq. (42), the qualitative validity of the simple theory is indicated by the way in which the frequencies of the overtone modes may be arranged in three series and the modal indices assigned, as shown in Fig. 5.7.

The abscissa of Fig. 5.7 is the frequency range between 3.10 and 3.75 MHz. Within this range the crystal unit has 13 modes of vibration, including the 101, or fundamental, mode. The first overtone mode is the 111 mode at 3.22 MHz. The second overtone is the 102 mode and it occurs at 3.27 MHz. All the modes can be arranged in series corresponding to k = 1, 2, 3, . . . and values of $m = 0, 1, 2, \ldots$. A plot of χ^2_{mk} shows a similar arrangement.

When the plate vibrates in the $n01$ mode, all points on the surface move in phase, so the surface charge has the same sign at all points. This condition is illustrated in the drawing in Fig. 5.8. In filter applications the $n11$ mode is the most troublesome because it is the first mode above the fundamental, or desired, mode. The index

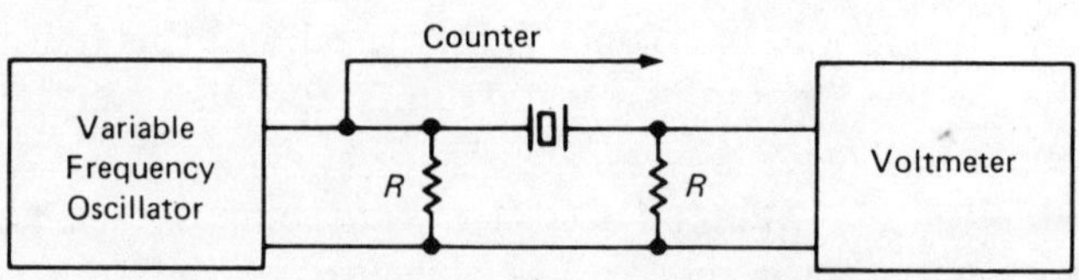

Fig. 5.6. Passive network for observing the frequencies of the modes of vibration of a quartz crystal unit.

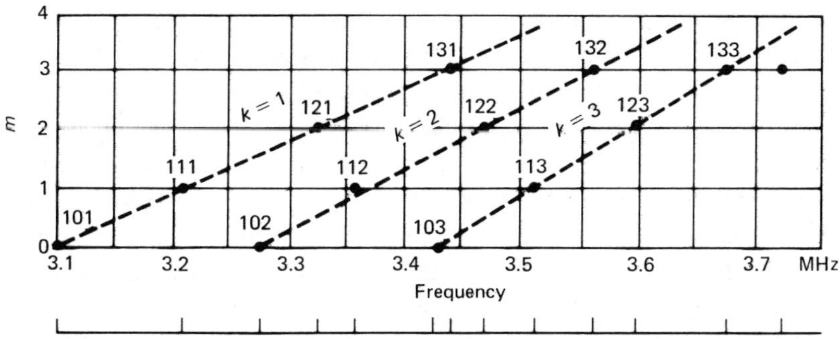

Fig. 5.7. Overtone spectrum of a 3.1-MHz AT-cut plate (experimental).

$m = 1$ indicates that one nodal line exists along a diameter of the plate. The polarization charges on opposite sides of the line have opposite signs, as indicated in Fig. 5.8. The index $m = 2$ indicates that two radial nodal lines exist, etc. The index $k = 1$ indicates that only one circular nodal line exists, i.e., the one surrounding the vibrating area. The index $k = 2$ indicates that two concentric nodal lines exist, etc. The electric charge developed on the two sides of any nodal line must be different, since the phase of the strain is opposite on the two sides.

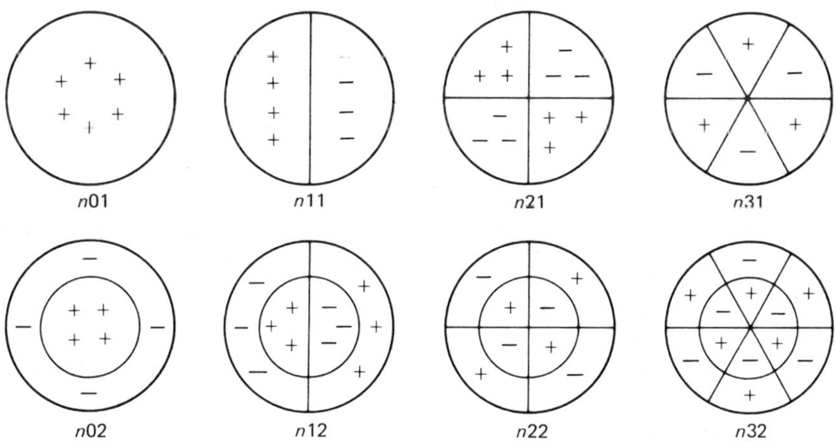

Fig. 5.8. Relative phase and charge on the surface of a thickness-shear plate vibrating in different modes *nmk*.

If the quartz plate and electrodes are perfectly symmetric with respect to the nodal line, then the $n11$ mode should not be excited, since the polarization charges would exactly cancel. The same thing is true for all modes having values of m other than $m = 0$, because the center of the plate is a nodal point with the charges developed symmetrically around it (neglecting the effects of anisotropy).

If the plate and electrodes are perfectly symmetrical and concentric with the vibrating area, then only modes having the modal indices $n0k$ could be excited. The first troublesome mode would then be the $n02$ mode in which two circular nodal lines are present. The charge inside the inner nodal line is opposite in sign to that developed on the surface between the two nodal lines. Conceivably it might be possible to suppress the $n02$ mode without losing the $n01$ mode by adjusting the electrode size so that the charges developed in the two zones would exactly cancel. Of course, the problem is greatly complicated by the anisotropy of the quartz.

SUMMARY

In this chapter we have tried to show that the vibration of a quartz piezoid is not unlike that of any other vibrating solid. The principle difference is the manner in which the vibrations are excited. The piezoelectric effect provides a particularly convenient method for setting the mechanical system into resonance. The overtone modes, some harmonic and some inharmonic, are characteristic of all vibrating bodies. There is nothing "spurious" about them, that is, the term should not imply that they are in some way unnatural or illegitimate.

We should recall, too, that we have been discussing idealized situations for the sake of simplicity. The complete problem of the vibrating piezoid including all the effects of anisotropy and with realistic boundary conditions is practically insoluble. We have neglected damping and driving forces which further complicate the problem. However, from a discussion of the idealized problem, it is possible to see the reasons for many of the phenomena which are observed and to suggest steps which may be taken to achieve desired design goals.

The idealized drawings of Fig. 5.8 must be understood to be illustrative but not of all features of the vibration. Nevertheless, the strain and charges on the surfaces of actual resonators do exhibit nodal lines as shown by dust patterns and other methods. Some overtone modes can be suppressed by careful attention to design of the quartz plate and to the deposition of the electrodes. It is apparent that the most meticulous care must be given to the geometry of the plate and to the symmetry of the electrodes in order to provide resonators which are suitably free of undesired modes for filter applications. It is clear that this can be done more exactly with round than with rectangular plates.

The material presented in this chapter should indicate that the design of a quartz resonator, taking into account, damping, anisotropy, overtone modes, mechanical imperfections, and coupled modes of vibration is not a simple process. The fabrication of a piezoid to a given design requires meticulous attention to details and care in every step of the process in order to realize the device as the designer intended.

The inharmonic overtone modes of thickness-shear resonators and the problem which they present to the designer of high-frequency crystal units for filter applications are discussed further in Chap. 10. The several measures which may be taken to minimize their effects including, particularly, contouring and energy trapping are treated in some detail.

In conclusion it may be worthwhile to describe a simple method for investigating the modes of motion of the vibrating piezoid. The piezoid is excited in one of its resonant modes by means of a self-excited oscillator or driven by means of an oscillator, as shown in Fig. 5.6. While it is vibrating, the surface is probed with a sharp probe made from a sliver of polyethylene or some similar material. Probing the surface at nodal points and along nodal lines has little or no effect on the damping, whereas touching the surface at an antinodal point has maximum effect. In this way a map may be made of the vibrating surface showing the nodes and antinodes of the mode of vibration. Much useful information about the nature of the vibration may be obtained in this way.

6
EQUIVALENT CIRCUIT OF THE QUARTZ RESONATOR

INTRODUCTION

From the point of view of the crystal engineer, the quartz crystal unit is a carefully prepared geometric solid, excited through the piezoelectric effect to vibrate at one of its normal resonance modes. To the circuit engineer, however, it is a two-terminal passive network having highly desirable characteristics, not realizable in any other way. The crystal resonator is really an electromechanical transducer, i.e., a device for converting electric to mechanical energy, and vice versa. Other examples are loudspeakers, microphones, electrically driven tuning forks, and magnetostriction resonators.

Several years before the discovery of the piezoelectric resonator, Dye and Butterworth showed that any mechanically vibrating system driven by means of an electromagnetic field could be represented by the electric network of Fig. 6.1*a* and that one driven by means of an electrostatic field could be represented by the network of Fig. 6.1*b*. The relationship between the parameters of the electric network and the physical properties of the piezoid were developed by Van Dyke, Butterworth, and others in the mid-twenties after the discovery of the piezoelectric resonator.[1]

In this chapter we attempt to show, in the simplest possible way, how the resistance, inductance, and capacitances of the equivalent

---

[1] D. W. Dye, *Proc. Phys. Soc.*, **38**: 399–457 (1926); S. Butterworth, *Proc. Phys. Soc.*, **27**: 410–424 (1915); K. S. Van Dyke, *Proc. I.R.E.*, **16**: 742–764 (1928).

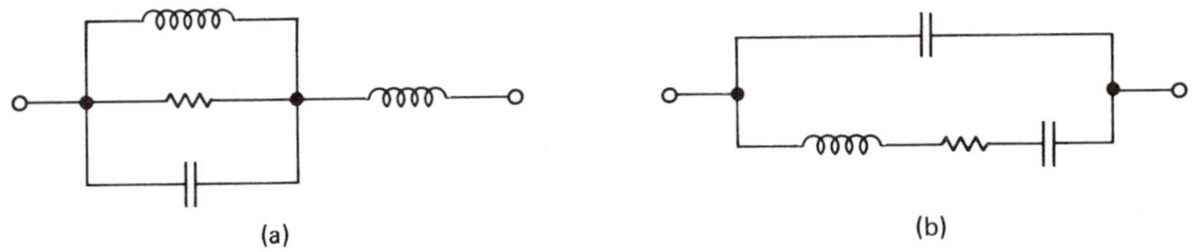

Fig. 6.1. Equivalent circuits of electromechanical transducers. (*a*) **Electromagnetic transducer;** (*b*) **electrostatic transducer.**

circuit are related to the dimensions, density, and elastic, dielectric, and piezoelectric constants of the piezoid. The equivalent circuit which we derive is an "exact" equivalent circuit, meaning that the parameters are independent of frequency, at least in the neighborhood of the resonant frequency and so long as only one mode of vibration is present. Simpler "limited" equivalent circuits may be used to represent the crystal unit at a single frequency.

In order to simplify the problem as much as possible we consider an infinite plate, vibrating in a thickness mode, thereby limiting the problem to one dimension. We assume a single mode of vibration and assume that all effects are linear, i.e., independent of amplitude. Other simplifications will be introduced at the appropriate time.

THE QUARTZ RESONATOR AS A VIBRATING CAPACITOR[2]

A piece of quartz or other piezoelectric crystal placed in an electric field will, in general, experience strains which are proportional to the strength of the field and change signs if the direction of the field is reversed. At the same time the quartz serves as the dielectric in a capacitor. The crystal unit may thus be thought of as a capacitor in which the dielectric vibrates in response to the applied field.

Let the piece of quartz be in the shape of a thin plate cut with its thickness in the Y-direction. Imagine electrodes to be plated on the surfaces. A difference of potential is established between the electrodes. Two types of strain exist due to the resulting electric field. One strain may be described as a pure shear in the XZ-plane, the other as a pure shear in the XY-plane. The first, or face-shear, strain is used in the fabrication of the low-frequency resonators –

[2] The reader may find it useful to review App. I at this point.

the CT- and DT-cut plates. The second type of strain is used in the *Y*-cut family, which includes the AT- and BT-cut plates. In this chapter we are exclusively concerned with the high-frequency or thickness-shear strain (although it should not be forgotten that the face-shear strain is also present).

Since the mechanical strain is proportional to and therefore reverses sign with the applied field, an alternating field causes an alternating strain. Unless the applied frequency is equal to that of a resonant frequency of the plate, the strain is small; but if the frequency of the applied field is nearly equal to that of the resonant frequency, the strain may become very large. In fact, it is quite easy to cause the strain to exceed the elastic limit and to break the quartz.

The amount of energy required to overcome frictional losses and to maintain the plate in vibration at constant amplitude is very small, so the normal resonance frequency of the plate is influenced very little by the driving forces applied through the piezoelectric effect. With appropriate connections to an amplifier, the mechanical vibration of the quartz resonator can be made to control the frequency of oscillation in the electric circuit. It can also be used as a passive component in an electric-filter network.

At frequencies far from the resonant frequency, the quartz plate is equivalent to a simple parallel-plate capacitor having a capacitance given by

$$C_0 = k\,\epsilon_0\,\frac{A}{e}$$

in mks units, where A is the area of the electrodes and e is the thickness of the plate. The dielectric constant k must be determined for the direction of the field, and unless the plate is infinitely large, correction must be made for fringing of the field. For the moment, we will not distinguish between the clamped and unclamped dielectric constant, although it should be pointed out that a piezoelectric crystal has two dielectric constants depending on whether the crystal is clamped or not. If the crystal is constrained in such a way that the strain is zero, the effective dielectric constant is not the same as if the crystal is completely free with the applied stress equal to zero. Fortunately the two values are not very different.

The electric conductivity of quartz is extremely low at ordinary temperatures, so that electrically the quartz plate and its electrodes may be represented by a pure capacitance. At frequencies other than those at which mechanical resonance occurs, the device is simply a capacitor and the current into it is simply the charging current, which is the displacement current due to the polarization of the dielectric just as in any other capacitor. This current leads the applied voltage by 90°; therefore the impedance is given by[3]

$$Z = -\frac{j}{\omega C_0}$$

where, as usual, $j = \sqrt{-1}$ and ω is the angular frequency $2\pi f$.

At frequencies near that of one of the normal modes of vibration of the piezoid, the strain is greatly increased. The strain associated with the vibration results in further polarization through the piezoelectric effect. The current flowing into the resonator then includes two components – the dielectric displacement current which was discussed earlier and the piezoelectric displacement current. The piezoelectric displacement current may be hundreds of times greater than the dielectric displacement current. The resonator current is the vector sum of the two.

The component due to the dielectric displacement current is 90° out of phase with the applied voltage, as in any perfect capacitor; and insofar as this component of the current is concerned, the resonator may be represented as an ordinary capacitor. The component of current which is due to the piezoelectric strain may either lead or lag the applied voltage, depending on whether the applied frequency is greater or less than that of the resonant frequency of the resonator. If the applied frequency is exactly equal to the mechanical resonant frequency, the strain and therefore the piezoelectric displacement current is in phase with the applied voltage, just as in the case of a resistor. The resonator then appears to be a pure resistance in parallel with a capacitor, as shown in Fig. 6.2. The current into the resonator is therefore the vector sum of the currents i_0 and i.

At frequencies above the resonant frequency, the piezoelectric current lags behind the applied voltage, i.e., the device behaves like an inductance and the current is the vector sum of i_0 and i', as shown in Fig. 6.3*a*. At frequencies below the resonant frequency, the pi-

[3] In this, and following chapters, extensive use will be made of complex notation. Familiarity with complex numbers is a prerequisite. See App. II.

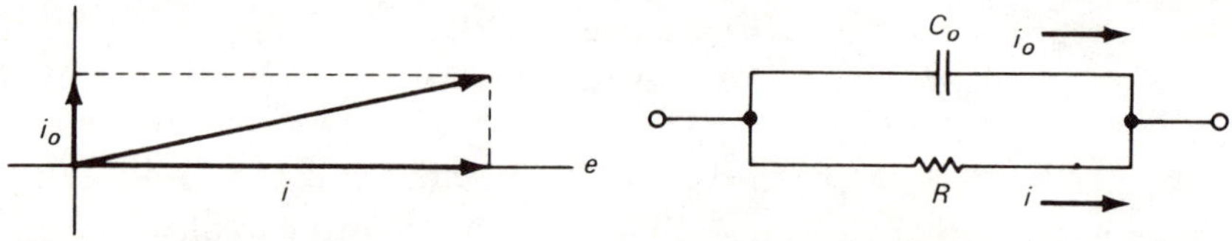

Fig. 6.2. The equivalent circuit of a quartz resonator at the frequency of mechanical resonance (currents not to scale).

ezoelectric current into the resonator leads the applied voltage and the device behaves as if it were a capacitor and the current is the vector sum of i_0 and i'', as shown in Fig. 6.3*b*.

If we consider a series circuit consisting of an inductance, a capacitance, and a resistance, we see that the current varies with frequency in the same way that the piezoelectric displacement current varies with frequency in the resonator. In both cases the current and voltage are in phase at the resonant frequencies. At higher frequencies the inductive reactance predominates and the current lags behind the voltage. At frequencies below the resonant frequency the capacitive reactance is predominant and the voltage lags behind the current. In each case the current into the resonator is the vector sum of the two currents, as indicated in Fig. 6.3.

DERIVATION OF THE EQUIVALENT CIRCUIT

We have just shown qualitatively that the piezoelectric resonator is equivalent electrically to a circuit consisting of a capacitor C_0 in parallel with the series circuit L, C, and R, as shown in Fig. 6.4.

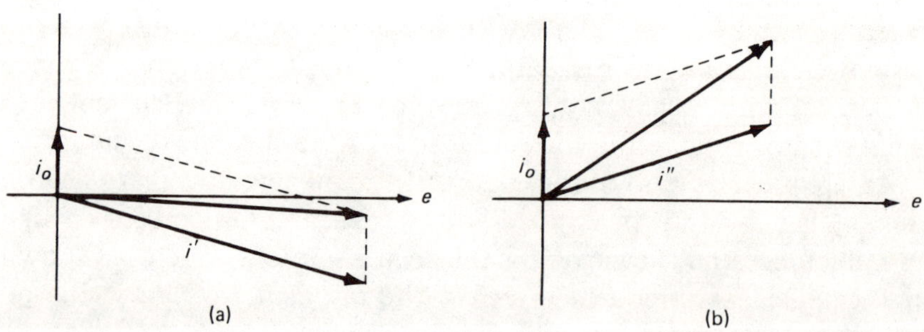

Fig. 6.3. Dielectric and piezoelectic currents at frequencies (*a*) below and (*b*) above the frequency of resonance.

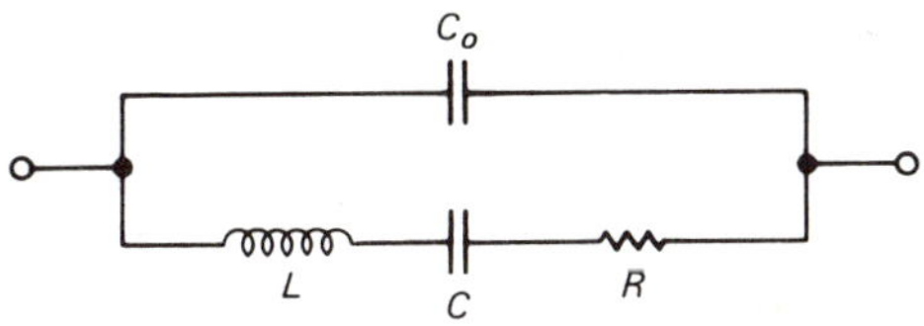

Fig. 6.4. The equivalent circuit of the piezoelectic resonator.

The capacitance $C_0 = k\epsilon_0 A/e$, where A is the electrode area, e is the thickness, and k is the applicable dielectric constant. The values of L, C, and R are much more difficult to compute, depending as they do upon the piezoelectric effect for the coupling between the mechanical strain in the resonator and the parameters of the equivalent circuit. The complete treatment of the problem including finite-boundary conditions, anisotropy, and nonlinear effects would be prohibitively complex, but a simplified treatment which takes into account only the most important factors yields results which are of considerable value in the design of quartz crystal units. For the complete treatment, the reader is referred to Cady's* *Piezoelectricity* (Chap. 14).

The steps which we follow in deriving the expressions for the parameters of the equivalent circuit are the following:

1. Solve the equation of motion, including the effects of damping, assuming that the damping forces are proportional to the particle velocity.
2. Apply appropriate boundary conditions to determine the displacement at all points in the vibrating plate.
3. Determine the strain by differentiating the displacement equation.
4. Determine the polarization by multiplying the strain by the appropriate piezoelectric constant.
5. Compute the space charge density by taking the negative of the gradient of the polarization,

$$q_v(y) = -\frac{\partial P_y}{\partial_y}$$

*W.G. Cady, *Piezoelectricity,* McGraw-Hill (1946) Chap. 14.

6. Use Poisson's equation:

$$V'' = \frac{q_v}{k\epsilon_0}$$

to obtain the potential at each point in the plate. (See symbols below.)

7. Evaluate the potential function at the surface of the plate.
8. Compute the field intensity at the surface from the relation

$$E = -\frac{\partial V}{\partial y}$$

(See symbols below.)

9. Determine the current density from

$$i_s = k\epsilon_0 \dot{E}$$

10. Divide the piezoelectric current by the applied voltage to obtain the electric admittance of the resonator, thereby obtaining the parameters of the equivalent circuit.

In order to keep the problem as simple as possible, we make the following assumptions.

1. The piezoid is an infinitely large, thin plate. This reduces the problem to one dimension.
2. The displacement is considered to be a wave traveling in the thickness direction. (However, no distinction between transverse and longitudinal waves is required except in the constants.)
3. Anistropy effects are neglected.
4. The electrodes are deposited directly on the plates, have negligible mass, and are perfectly conducting. The complication of an air gap is unnecessary, since few, if any, such devices are produced.
5. A single mode of vibration with no coupling to other modes is assumed.
6. The thickness of the plate is taken to be in the y direction, with the origin of coordinates at the center of the plate. The thickness of the plate is e and the surfaces are therefore at $y = \pm\frac{1}{2}e$.
7. The strain is considered to be constant in a plane parallel to the plane $y = 0$.

The following symbols are required.

A = Area of the plate
e = Thickness of the plate
c = Appropriate elastic constant
d = Piezoelectric strain constant
D = Electric displacement
$E = E(y, t)$ = Electric-field intensity, $\dot{E} = \partial E/\partial t$
f = Frequency (Hz, kHz, MHz, etc.)
i_s = Current density
k = Dielectric constant
P = Polarization
q_v = Electric-charge density
r = Damping coefficient
t = Time
v = Wave velocity
$V = V(y,t)$ = Potential
V_m = Amplitude of driving voltage
ϵ = Piezoelectric stress constant
ϵ_0 = Susceptibility of space = $8.85 \times 10^{-12}\ \mathrm{C^2/N \cdot m^2}$ (or $\mathrm{C/V \cdot m}$)
γ = Propagation constant
ρ = Density of quartz = $2.65\ \mathrm{g/cm^3} = 2650\ \mathrm{kg/m^3}$
ψ = Mechanical displacement
$\psi' = \dfrac{\partial \psi}{\partial y}$ = Strain
$\omega = 2\pi f$ = Angular frequency (rad/sec)

$$\psi'' = \frac{\partial^2 \psi}{\partial y^2} \qquad V'' = \frac{\partial^2 V}{\partial y^2}$$

$$\dot{\psi} = \frac{\partial \psi}{\partial t} \qquad V' = \frac{\partial V}{\partial y}$$

$$\ddot{\psi} = \frac{\partial^2 \psi}{\partial t^2} \qquad \dot{V} = \frac{\partial V}{\partial t}$$

The equation of motion of each point in the plate, including the effect of damping, is

$$\rho \ddot{\psi} - c\psi'' + r\dot{\psi} = 0 \tag{44}$$

The assumption that the damping force is proportional to the velocity is justified by the results (and the not inconsiderable fact that it results in an equation which can be solved). Little theoretical justification is available for the assumption; it may be that some of the observed variations of frequency with amplitude are due to nonlinear damping effects. It seems much more likely, however, that other effects are responsible.

The generalized stiffness constant c is used when considering the vibration of plates, whereas the compliance constant s is used in the problem of a vibrating bar. This distinction is necessary because the longitudinal vibration of a bar involves only the extensional stresses, the lateral stresses being negligible. In a thin plate, the lateral strains are zero; the lateral stresses are, in general, unknown.

The solution of Eq. (44) may be found by the usual method of assuming that the displacement ψ is the product of a function $A(y)$ and a function $\exp(j\omega t)$, i.e.,

$$\psi = A(y)\exp(j\omega t) \tag{45}$$

Substitution in Eq. (44) leads to the following solution.

$$\psi = [C_1 \sin(\gamma y) + C_2 \cos(\gamma y)] \exp(j\omega t) \tag{46}$$

where γ is the complex quantity

$$\gamma = \left[\frac{\omega^2}{v^2} - \frac{j\omega r}{\rho v^2}\right]^{\frac{1}{2}} \tag{47}$$

and the wave velocity $v = (c/\rho)^{\frac{1}{2}}$.

In case the damping coefficient $r = 0$, then $\gamma = \omega/v$ and the solution is the same as the one obtained earlier. [See Eq. (30)].

From the boundary condition that $\psi = 0$ at the center of the plate where $y = 0$, it follows that $C_2 = 0$, leaving

$$\psi = C_1 \sin(\gamma y)\exp(j\omega t) \tag{48}$$

To evaluate the constant C_1 we make use of the fact that the stress at the surface is due to the electric field alone; the mechanical stress due to internal tractive forces is zero. This is not the case in the interior of the plate, where stress (and strain) results from the combined mechanical and piezoelectric stresses. At the surface where $y = +\frac{1}{2}e$, the strain

$$\psi' = dE = d\,\frac{V_m}{e}\,\exp(j\omega t)$$

where $V_m \exp(j\omega t)$ is the applied voltage, assumed to be sinusoidal.

Differentiating Eq. (48) and evaluating at $y = \frac{1}{2}e$, we have

$$\psi' = \gamma C_1 \cos(\tfrac{1}{2}\gamma e)\exp(j\omega t) \tag{49}$$

Equating the last two expressions for ψ' and solving for C_1 gives

$$C_1 = \frac{d\,V_m}{\gamma\, e \cos(\frac{1}{2}\gamma e)}$$

The solution of Eq. (44) consistent with the boundary conditions including the driving voltage $V_m \exp(j\omega t)$ is

$$\psi = \frac{d\,V_m}{\gamma e \cos(\frac{1}{2}\gamma e)}\sin(\gamma y)\exp(j\omega t) \tag{50}$$

The polarization $P = \epsilon\psi'$.

The charge density

$$q_v = -\frac{\partial P}{\partial y} = -\epsilon\psi''$$

Differentiating Eq. (50) twice with respect to y gives

$$\psi'' = -\frac{\gamma V_m d}{e\cos(\frac{1}{2}\gamma e)}\sin(\gamma y)\exp(j\omega t)$$

From Poisson's equation (See App. I),

$$V'' = -\frac{q_v}{k\epsilon_0} = \frac{\epsilon}{k\epsilon_0}\psi''$$

so that

$$V'' = -\frac{\epsilon\gamma d\,V_m}{k\epsilon_0\, e\cos(\frac{1}{2}\gamma e)}\sin(\gamma y)\exp(j\omega t) \tag{51}$$

Integrating once,

$$V' = \frac{\epsilon d V_m}{k \epsilon_0 e \cos(\frac{1}{2} \gamma e)} \cos(\gamma y) \exp(j\omega t) + C_3$$

and integrating again,

$$V = \frac{\epsilon d V_m}{\gamma k \epsilon_0 e \cos(\frac{1}{2} \gamma e)} \sin(\gamma y) \exp(j\omega t) + C_3 y + C_4 \qquad (52)$$

From the boundary condition that $V = 0$ when $y = 0$, we have that $C_4 = 0$. The other boundary condition is

$$V = \tfrac{1}{2} V_m \exp(j\omega t) \qquad \text{at } y = +\tfrac{1}{2} e$$

from which

$$\tfrac{1}{2} V_m \exp(j\omega t) = \frac{\epsilon d \sin(\frac{1}{2} \gamma e)}{k \epsilon_0 e \cos(\frac{1}{2} \gamma e)} V_m \exp(j\omega t) + C_3 (\tfrac{1}{2} e)$$

and

$$C_3 = \left[\frac{1}{e} - \frac{2 \epsilon d}{\gamma k \epsilon_0 e^2} \tan(\tfrac{1}{2} \gamma e)\right] V_m \exp(j\omega t)$$

Substituting in Eq. (52) gives the expression for the potential $V(y)$ within the plate.

$$V = \left[\frac{\epsilon d \sin(\gamma y)}{\gamma k \epsilon_0 e \cos(\frac{1}{2} \gamma e)} + \frac{y}{e} - \frac{2 \epsilon d}{\gamma k \epsilon_0 e^2} \tan(\tfrac{1}{2} \gamma e) y\right] V_m \exp(j\omega t)$$

The electric field

$$E(y) = -\frac{\partial V}{\partial y}$$

and thus

$$E = \left[\frac{\epsilon d \cos(\gamma y)}{k \epsilon_0 e \cos(\frac{1}{2} \gamma e)} + \frac{1}{e} + \frac{2 \epsilon d}{\gamma k \epsilon_0 e^2} \tan(\tfrac{1}{2} \gamma e)\right] V_m \exp(j\omega t)$$

At the surface $y = +\frac{1}{2}e$:

$$E_{y=\frac{1}{2}e} = \left[\frac{1}{e} + \frac{\epsilon d}{k\epsilon_0 e} + \frac{2\epsilon d}{\gamma k \epsilon_0 e^2}\tan\left(\tfrac{1}{2}\gamma e\right)\right] V_m \exp(j\omega t)$$

The current density at the surface of the resonator is

$$i_s = \frac{\partial D}{\partial t}$$

or since $D = k\epsilon_0 E$,

$$i_s = k\epsilon_0 \frac{\partial E}{\partial t}$$

and therefore the current density at the surface of the resonator is

$$i_s = j\omega\left[\frac{k\epsilon_0}{e} + \frac{\epsilon d}{e} + \frac{2\epsilon d}{\gamma e^2}\tan\left(\tfrac{1}{2}\gamma e\right)\right] V_m \exp(j\omega t)$$

The total current into the resonator, assuming the current density (current per unit area) to be constant over the surface, is

$$i = \left[j\omega A\left(\frac{k\epsilon_0}{e} + \frac{\epsilon d}{e}\right) + j\omega A\,\frac{2\epsilon d}{\gamma e^2}\tan\left(\tfrac{1}{2}\gamma e\right)\right] V_m \exp(j\omega t) \quad (53)$$

The two terms inside the bracket of Eq. (53) have the dimensions of an admittance, since each is a current divided by a voltage. The first term

$$j\omega A\left[\frac{k\epsilon_0}{e} + \frac{\epsilon d}{e}\right]$$

is the admittance of a capacitor having capacitance

$$C_0 = (k\epsilon_0 + \epsilon d)\frac{A}{e}$$

If the piezoelectric constants were zero or the plate completely clamped so that all mechanical strains were zero, then

$$C_0 = k\epsilon_0 \frac{A}{e}$$

which is the capacitance of an ordinary parallel-plate capacitor. The term ϵd is about 1 percent of $k\epsilon_0$ so that the capacitance of a parallel-plate capacitor having a mechanically free quartz dielectric would be reduced by about 1 percent by clamping the plate to suppress the strain.

The second term in Eq. (53) represents another admittance in parallel with C_0. This admittance is obviously complex, since the quantity γ is complex. Therefore, the quartz plate with its electrodes is electrically equivalent to a capacitor in parallel with a complex admittance whose value depends upon the frequency. It is now necessary to investigate the nature of this admittance.

From Eq. (47) γ represents the quantity

$$\gamma = \left(\frac{\omega^2}{v^2} - \frac{j\omega r}{\rho v^2}\right)^{\frac{1}{2}}$$

The second term in the expression for γ is very small, so we may write with a high degree of approximation

$$\gamma = \frac{\omega}{v} - j\frac{r}{2\rho v} \tag{54}$$

For convenience in notation, we let $r/2\rho v = \beta$ so that Eq. (54) may be written

$$\gamma = \frac{\omega}{v} - j\beta$$

From the identity

$$\tan(x - jy) = \frac{\sin(2x) - j\sinh(2y)}{\cos(2x) + \cosh(2y)}$$

$$\tan\tfrac{1}{2}\gamma e = \tan\left[\frac{\omega e}{2v} - j\frac{\beta e}{2}\right] = \frac{\sin(\omega e/v) - j\sinh(\beta e)}{\cos(\omega e/v) + \cosh(\beta e)} \tag{55}$$

The angular frequency ω may be written as $\omega_0 - \Delta\omega$, where $\Delta\omega$ is always very small (almost always less than 0.1 percent) compared with ω or ω_0. This is because appreciable mechanical response occurs only when the applied frequency ω is very nearly equal to the resonant frequency ω_0.

The quantity $(\omega e/v)$ may be written

$$\frac{\omega e}{v} = \left(\frac{\omega_0 e}{v} - \frac{\Delta\omega e}{v}\right) = \left(\frac{2\pi f_0 e}{v} - \frac{\Delta\omega e}{v}\right) = \left(\pi - \frac{\Delta\omega e}{v}\right)$$

With this substitution Eq. (55) may be written

$$\tan\left(\tfrac{1}{2}\gamma e\right) = \frac{\sin(\pi - \Delta\omega e/v) - j\sinh(\beta e)}{\cos(\pi - \Delta\omega e/v) + \cosh(\beta e)}$$

which is equal to

$$\tan\left(\tfrac{1}{2}\gamma e\right) = \frac{\sin(\Delta\omega e/v) - j\sinh(\beta e)}{-\cos(\Delta\omega e/v) + \cosh(\beta e)} \tag{56}$$

But since $(\Delta\omega\, e/v)$ is a very small angle (equal to $\Delta\omega\pi/\omega_0$), it is permissible to make the following substitutions:

$\left(\frac{\Delta\omega e}{v}\right)$ for $\sin\left(\frac{\Delta\omega e}{v}\right)$

$1 - \frac{1}{2}\left(\frac{\Delta\omega e}{v}\right)$ for $\cos\left(\frac{\Delta\omega e}{v}\right)$

(βe) for $\sinh(\beta e)$

$1 + \frac{1}{2}(\beta e)^2$ for $\cosh(\beta e)$

With these approximations and a considerable amount of simple algebra, Eq. (56) may be written as

$$\tan\left(\tfrac{1}{2}\gamma e\right) = \frac{2v}{e}\,\frac{\Delta\omega - j\alpha}{\Delta\omega^2 + \alpha^2} \qquad \text{where } \alpha = \frac{r}{2\rho} \tag{57}$$

The second term in the bracket of Eq. (53) may now be written as

$$Y_m = \frac{4j\,v\,\omega A\,\epsilon d}{\gamma e^3}\;\frac{\Delta\omega - j\,\alpha}{\Delta\omega^2 + \alpha^2}$$

Using the approximation,

$$\gamma = \omega_0/v = \pi/e,$$

$$Y_m = \frac{4j\omega\,\omega_0\,A\,\epsilon d}{\pi^2 e}\;\frac{\Delta\omega - j\,\alpha}{\Delta\omega^2 + \alpha^2}$$

The reciprocal of the admittance Y_m is the impedance Z_m, which is

$$Z_m = \frac{\pi^2 e}{4\,\omega\,\omega_0\,A\epsilon d}\,(\alpha - j\,\Delta\omega)$$

The resistance part of Z_m is

$$R_m = \frac{\pi^2\,er}{8\,\omega_0^2 A\,\epsilon d\,\rho} \tag{58}$$

Equation (58) may be simplified by expressing d in terms of ϵ and the elastic constant c. In matrix notation

$$\begin{aligned} x &= dE && \text{from Eq. (15)}\\ -X &= \epsilon E && \text{from Eq. (20)}\\ -X &= cx && \text{from Eq. (13)}\\ -x &= c^{-1}X. \end{aligned}$$

From the four equations above we have $x = c^{-1}\epsilon E = dE$ and therefore

$$d = c^{-1}\epsilon$$

In a simple one-dimensional case c^{-1} is equivalent to $1/c$, so that we have $d = \epsilon/c$. Substituting for d and recalling that

$$f_0 = \frac{1}{2e}\sqrt{\frac{c}{\rho}}$$

we obtain

$$R_m = \frac{\pi^2 erc}{8\,\omega_0^2\,A\,\epsilon^2\rho}$$

$$R_m = \frac{e^3 r}{8A\epsilon^2} \tag{59}$$

The expression above for R_m shows that the motional resistance in the equivalent network is directly proportional to the damping constant r and to the cube of the thickness of the plate. R_m is inversely proportional to the area A and to the square of the piezoelectric stress constant ϵ. It is not possible to compute the numerical value of R, since no direct measurement of the value of r is possible. It must be recalled that r is assumed to include all dissipative forces and that they are proportional to the particle velocity in the quartz. Other sources of damping, in addition to internal dissipation, are air damping, surface friction, and mounting losses. In most cases these are greater than the internal losses and it is unlikely that each is exactly proportional to the particle velocity. Possibly a part of the variation of resistance with amplitude which is sometimes observed may be due to nonlinear damping effects. The expression for R is valid even though r should be frequency-dependent; and if the dependence were known, the variation of the value of R with frequency could be calculated.

Despite its semiquantitative nature, Eq. (59) shows why the equivalent resistance of a low-frequency plate is so high compared with that of a high-frequency plate of the same area.

The reactive part of Z_m is given by

$$X_m = -\frac{\Delta\omega\pi^2 e}{4\omega_0^2 d \epsilon A}$$

Eliminating d as was done in the derivation of R gives

$$X_m = -\frac{\Delta\omega\rho e^3}{4A\epsilon^2}$$

Thus X_m varies directly with $-\Delta\omega$, the difference between the frequency of resonance and the operating frequency. This is precisely the behavior of a circuit consisting of an inductance and a capacitance in series. This may be shown in the following way.

The reactance of a circuit consisting of L and C in series is

$$X = \omega L - \frac{1}{\omega C} = \frac{\omega^2 LC - 1}{\omega C} = \frac{\omega^2 LC - \omega_0^2 LC}{\omega C}$$

$$= \frac{(\omega + \omega_0)(\omega - \omega_0)}{\omega} L \doteq -\frac{2\omega\Delta\omega}{\omega} L = -2\Delta\omega L\dagger$$

---

†Note that $\Delta\omega = (\omega_0 - \omega)$ (see page 94).

Therefore the reactance X_m behaves as if it were an inductance L having the value

$$L = \frac{\rho e^3}{8A\,\epsilon^2} \tag{60}$$

in series with a capacitance having the value

$$C = \frac{8A\,\epsilon^2}{\pi^2\,e\,c} \tag{61}$$

We have shown that a quartz plate with its electrodes is the equivalent of the electric circuit shown in Fig. 6.5, in which the parameters have the following values.

$$C_0 = k\,\epsilon_0\,\frac{A}{e}$$

$$R_m = R = \frac{e^3 r}{8A\,\epsilon^2}$$

$$L_m = L = \frac{e^3 \rho}{8A\,\epsilon^2}$$

$$C_m = C = \frac{8A\,\epsilon^2}{\pi^2\,e\,c}$$

In deriving the equations for R, L, and C it has been assumed that the strain distribution is uniform over the surface of the plate A. The strain over the surface of an actual resonator is, of course, not constant. The results obtained by the use of Eqs. (60) and (61) must be expected to be approximately valid and to differ from the measured values by factors of the order of unity. The equations do show how R, L, and C depend upon the physical properties of quartz and the dimensions of the plate and they therefore point the direction which must be taken to obtain desired variations in the parameters L and C.

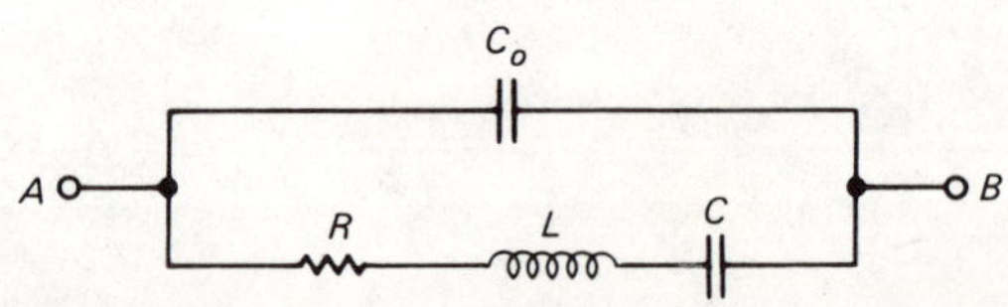

Fig. 6.5. The equivalent circuit of the piezoelectric resonator. The parameters R, L, and C are often called R_m, L_m, and C_m to call attention to the fact that they are associated with the motion of the plate.

In order to calculate the numerical values of L and C one must make a more elaborate calculation, taking into account the boundary conditions on a finite plate. Such a calculation is difficult. Alternatively, one may make some assumption concerning the distribution of the strain over the surface. Cady assumed the strain to be sinusoidal in the case of rectangular plates. The result agrees fairly well with measured values.

To investigate the consequences of the approximations made in deriving the expressions for C_0, C_m and L_m we may consider a certain 10.8-MHz unit made with an AT-cut piezoid having the following dimensions:

Diameter of blank	8.0×10^{-3} m	(315 mils)
Electrode diameter	2.5×10^{-3} m	(100 mils)
Blank thickness e	1.54×10^{-4} m	(6.06 mils)
Electrode area A	4.91×10^{-6} m^2	(0.010 in^2)
Angle of cut θ	$35\frac{1}{4}°$	

We must determine the value of the applicable piezoelectric coefficient ϵ'_{26} and the applicable stiffness coefficient c'_{66}. For the AT-cut these are

$$c'_{26} = (c_{14} s - \epsilon_{11} c)\, c$$

and $$c'_{66} = c_{44} s^2 + 2 c_{14}\, sc + c_{66}\, c^2$$

where $s = \sin 35\frac{1}{4}°$ and $c = \cos 35\frac{1}{4}°$. These values are

$$\epsilon'_{26} = 9.65 \times 10^{-2}\ \mathrm{C/m^2} \qquad \text{and} \qquad c'_{66} = 29.3 \times 10^{9}\ \mathrm{N/m^2}$$

Substituting these values in the expressions for C_0, C_m, and L_m and remembering to use mks units for all quantities, we obtain the theoretical values to be compared with the measured values with the following results:

	Theoretical	Measured
C_0	1.2 pF (plus holder capacitance)	2.2 pF (including holder)
C_m	8.2×10^{-15} F	6.6×10^{-15} F
L_m	0.026 H	0.033 H

The theoretical value of C_m is too large by about 25 percent and the theoretical value of L_m is too small by about the same amount. This is to be expected, since we have assumed that the strain is constant over the entire electrode area. This is impossible in a finite plate, of course. Nevertheless, the computed values are close enough to the actual measured values to be useful in predicting the parameters to be expected from a given design.

7
THE CRYSTAL UNIT AS A CIRCUIT COMPONENT

INTRODUCTION

In the previous chapter we have shown that the piezoelectric resonator is electrically equivalent to the network of Fig. 6.5. It is impossible, by any set of electrical measurements made at terminals A and B, to distinguish between a piezoelectric resonator and an electric circuit made up of these components. One might ask, if this be true, "What is the advantage of the quartz crystal unit over a conventional circuit?" The answer lies in the values which conventional components would have to provide in order to simulate the behavior of the crystal unit. The simulation of C_0 would present no problem, but the simulation of the other parameters is, in general, quite impossible. For example, the value of C in a typical 5-MHz AT-cut unit may be of the order of 0.01 pF. A capacitor of this size could scarcely be built, since its leads, if any, would probably have more capacitance than this. The value of L is of the order of 0.1 H and R is perhaps 10 Ω. A coil having an inductance of 0.1 H would ordinarily require many turns of wire on a ferromagnetic core. To construct such a coil with a resistance of only 10 Ω in a volume of less than 0.5 cm^3 is probably impossible.

The properties which give the quartz crystal unit its unique ability to control a frequency are its high Q, or quality, factor and its low κ, or electromechanical, coupling.

The Q of any resonant system is given by

$$Q = 2\pi \frac{\text{Energy stored per cycle}}{\text{Energy dissipated per cycle}}$$

The Q of a series-resonant electrical circuit is given by $Q = \omega L/R = 1/\omega RC$ and may have a value of a few hundred. The Q of a well-made tuning fork may be a few thousand, but the Q of a quartz crystal unit is usually not less than 10^5 and may be as high as 10^7.

The value of κ, the electromechanical coupling coefficient, is given by

$$\kappa = \sqrt{\frac{\text{Energy stored elastically}}{\text{Energy stored electrically}}}$$

The value of κ^2 for an AT-cut plate (see App. IV) is less than 1 percent. This means that the electric driving system is very loosely coupled to the mechanical system and that the former, therefore, has very little influence over the latter. But because of the high Q of quartz, even this small effect is sufficient to excite and maintain vibrations in the quartz piezoid, which is what makes it an effective transducer for controlling a frequency.

THE RESONANT FREQUENCIES

The admittance Y_{AB} between the terminals A and B of Fig. 6.5 is given by

$$Y_{AB} = \frac{1}{Z_{AB}} = \frac{1}{R + j(\omega L - 1/\omega C)} + j\omega C_0$$

$$= \frac{1 + j\omega R C_0 - \omega^2 L C_0 + C_0/C}{R + j(\omega L - 1/\omega C)}$$

$$Z_{AB} = \frac{R + j(\omega L - 1/\omega C)}{1 - \omega^2 L C_0 + C_0/C + j\omega R C_0}$$

Multiplying the numerator and denominator by the complex conjugate of the latter gives

$$Z_{AB} = \frac{R + j(\omega L - 1/\omega C - \omega^3 L^2 C_0 + 2\omega L C_0/C - C_0/\omega C^2 - \omega R^2 C_0)}{C_0^2/C^2 + 1 - 2\omega^2 L C_0 + 2C_0/C + \omega^2 R^2 C_0^2 - 2\omega^3 L C_0^2/C + \omega^4 L^2 C_0^2} \tag{62}$$

The condition for the circuit to be resonant is that the imaginary part of Eq. (62) should be equal to zero. Setting the imaginary part of Eq. (62) equal to zero and clearing of fractions we have

$$\omega^2 LC^2 - C - \omega^4 L^2 C^2 C_0 + 2\omega^2 LCC_0 - C_0 - \omega^2 R^2 C^2 C_0 = 0 \tag{63}$$

For most practical resonators and within the accuracy of many measurements, the value of the last term containing the resistance R is negligible compared with the other terms and may be omitted in calculating the resonant frequencies. Omitting this term and rearranging gives

$$\omega^4 L^2 C^2 C_0 - \omega^2 (LC^2 + 2LCC_0) + (C + C_0) = 0 \tag{64}$$

Solving Eq. (64) by means of the quadratic formula, we obtain

$$\omega^2 = \frac{2LCC_0}{2L^2C^2C_0} \qquad \text{and} \qquad \omega^2 = \frac{2LC^2 + 2LCC_0}{2L^2C^2C_0}$$

The first root gives the series resonant frequency and the second the parallel or antiresonant frequency. Denoting these frequencies by f_R and f_A, respectively, we have

$$\omega_R^2 = \frac{1}{LC} \qquad \text{or} \qquad f_R = \frac{1}{2\pi}\sqrt{\frac{1}{LC}} \tag{65}$$

$$\omega_A^2 = \frac{1}{LC} + \frac{1}{LC_0} \qquad \text{or} \qquad f_A = \frac{1}{2\pi}\sqrt{\frac{1}{LC} + \frac{1}{LC_0}} \tag{66a}$$

If a capacitance C_x is connected in parallel with C_0, the antiresonant frequency is given by

$$f_A = \frac{1}{2\pi}\sqrt{\frac{1}{LC} + \frac{1}{LC_t}} \tag{66b}$$

where $C_t = C_0 + C_x$.

From Eq. (65) it follows that the series-resonant frequency is independent of the capacitance C_0. If the term involving R had been included in Eq. (64), the value of f_R would have been modified, but only by an insignificant amount. The frequency of parallel resonance f_A (sometimes called the antiresonant frequency) does depend on C_0. Therefore, when the unit is operated at the parallel-resonant frequency, the frequency is determined by the sum of C_0 and all capacitances in parallel with the crystal unit. Thus, while f_R is unique, f_A depends not only on the crystal unit but also on the associated circuit. From Eqs. (65) and (66*a*) it is seen that f_A is higher than f_R. In a sense, a crystal unit has an antiresonant frequency when it is not connected to any external circuit or to an external circuit which has zero reactance but the unit is usually not employed in this way.

THE IMPEDANCES AT THE RESONANT FREQUENCIES

The impedance of the crystal unit at the resonant frequency f_R is found by substituting $\omega_R^2 = 1/LC$ for ω in Eq. (62). The imaginary part of the numerator is zero, so the remainder of Eq. (62) reduces to

$$Z_R = \frac{R}{1 + (\omega R C_0)^2} \doteq R \tag{67}$$

The approximation in Eq. (67) is justified since for a typical AT unit operated at a frequency of 10 MHz, R is typically about 10 Ω, while the quantity $(\omega R C_0)^2$ is of the order of 10^{-5}. To a high degree of approximation the impedance of a crystal unit at the series-resonant frequency is thus equal to R. This approximation is valid in most cases.

The impedance of the crystal unit at the parallel or antiresonant frequency is found by substituting $\omega_A^2 = 1/LC + 1/LC_0$ into Eq. (62). In this way the antiresonant impedance is found to be

$$Z_A = \frac{1}{\omega^2 C_0^2 R} \quad \text{(Exact)} \tag{68a}$$

Any capacitance C_x in parallel with the crystal unit must be added to the capacitance C_0 so that

$$Z_A = \frac{1}{\omega^2 C_t^2 R} \qquad \text{where } C_t = C_0 + C_x \tag{68b}$$

The quantity $1/\omega^2 C_t^2 R$ has been called the *performance index* and it has been used as a measure of the quality of the crystal unit. It is not generally used now. If no extraneous coupled mode exists, the impedance of the crystal unit Z_A can be predicted at any frequency within the resonance interval. If coupled modes do exist, the value of Z_A may vary considerably from the calculated value.

At the series-resonant frequency f_R, the impedance of the crystal unit is a pure resistance having a relatively low value. Typical values for AT-cut units in the frequency range from 5 to 10 MHz vary between 10 and 100 Ω.

The impedance at the parallel-resonant frequency is also a pure resistance, but the value is much higher and, of course, depends upon the capacitance across the crystal unit. Typically Z_A is of the order of 10^4 to 10^5 Ω.

THE PARAMETERS OF THE EQUIVALENT CIRCUIT

The values of the reactive components of the equivalent circuit can be determined from measurements of the series- and parallel-resonant frequencies. Subtracting Eq. (65) from Eq. (66*a*) we have

$$\omega_A^2 - \omega_R^2 = (\omega_A - \omega_R)(\omega_A + \omega_R) = \frac{1}{LC_t}$$

where $C_t = C_0 + C_x$ is the total capacitance across the crystal unit.

Since ω_A and ω_R are very nearly equal we may write

$$2\omega\Delta\omega = \frac{1}{LC_t}$$

from which

$$L = (8\pi^2 f\, C_t\, \Delta f)^{-1} \qquad \text{where } \Delta f = f_A - f_R \tag{69}$$

Using this result and Eq. (65) we have

$$C = \frac{2\,\Delta f\, C_t}{f} \tag{70}$$

The quality factor Q, which is defined as $\omega L/R$, is thus given by

$$Q = (4\pi R\, C_t\, \Delta f)^{-1} \tag{71}$$

MEASUREMENT OF THE CIRCUIT PARAMETERS C_0 AND R.

The static capacitance C_0 may be measured by any convenient bridge or substitution method provided that the measurement is made near, but not too near, the resonant frequency. The measurement frequency should be near the operating frequency in order to take into account the distributed capacitance and inductance of the holder, both of which are usually frequency-dependent. A frequency about 1 percent below the resonant frequency is to be preferred. It is essential that the measurement be made at a frequency which does not coincide with one of the many inharmonic overtones of the plate. The range immediately below the resonant frequency is least likely to coincide with one of these modes. It is advisable to make the measurement at two or more different frequencies to ensure the absence of extraneous responses. The Q-meter provides a simple substitution method for measuring C_0. The accuracy is adequate for most purposes.

The value of R is easily determined by a circuit of the type illustrated in Fig. 7.1 (or its solid-state counterpart), which was the predecessor of the CI-meter to be discussed later.

The circuit is a simple self-excited shunt-fed Colpitts oscillator with a socket in series with the inductance. The circuit will oscillate

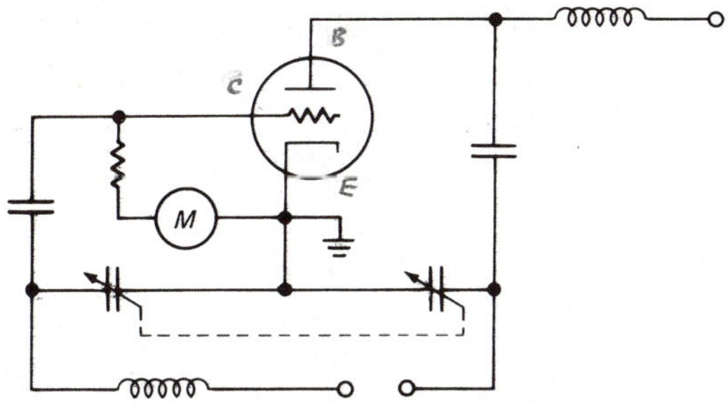

Fig. 7.1. Simple circuit for measuring R in the equivalent circuit.

if a resistor is plugged into the socket, provided the resistance is not too great. In practice a nonreactive resistor is inserted in the socket and the frequency is adjusted to be approximately equal to the resonant frequency of the crystal unit to be measured. The resistor is then replaced by the crystal unit and the frequency and grid current are noted. The crystal unit is then replaced by a resistor which results in the same grid current, and the tuning is adjusted until the same frequency and grid current are obtained with the resistor and the crystal unit. Under this condition, the crystal unit must be presenting the same reactance and resistance to the oscillator as the resistor and it is therefore operating at its series-resonant frequency. Thus both R and f_R are determined.

MEASUREMENT OF L AND C

The values of the motional parameters L and C are determined by measuring the frequencies of resonance and antiresonance and using Eqs. (69) and (70). We have just seen how the resonant frequency can be determined, but the measurement of the antiresonant frequency presents a more difficult problem.

From Eq. (66*b*) we see that the frequency f_A depends on the capacitance C_x across the crystal unit. Both C_x and f_A must be known to calculate L and C. When the crystal unit is used as an active element in an oscillator such as the Pierce or Miller circuits, the effective capacitance presented to the crystal unit is unknown and not readily measurable, because it depends not only upon the socket and wiring capacitances but also upon the transit time, Miller effect, electrode or junction capacitance, etc.

Since it is scarcely possible to determine C_x when the crystal unit is operated as the active element in an oscillator, it is usually operated as a passive element in *series* with the load capacitance C_x. The possibility of using this method depends on the result that the frequencies of zero reactance are the same in both cases, a fact that has resulted in considerable discussion and no little confusion since the CI-meter was invented in the mid-1940s. Before continuing with the discussion of methods of determining L and C we digress to show that the frequency of antiresonance given by Eq. (66*b*) is the same as the frequency of resonance of the network consisting of the crystal unit in series with the load capacitor C_x.

THE CRYSTAL UNIT IN SERIES WITH A CAPACITOR

In this section we derive the expressions for the frequencies of zero reactance and the corresponding impedances when the crystal unit is operated in series with a capacitor. Figure 7.2 shows the equivalent circuit of the crystal unit with a load capacitor C_x in parallel with the crystal unit (*a*) and in series with it (*b*).

The impedance between terminals A and B of the circuit of Fig. 7.2*b* is given by

$$Z_{AB} = \frac{R + j(\omega L - 1/\omega C)}{1 - \omega^2 LC_0 + C_0/C + j\omega RC_0} + \frac{1}{j\omega C_x}$$

$$= \frac{j\omega RC_x - \omega^2 LC_x + C_x/C + 1 - \omega^2 LC_0 + C_0/C + j\omega RC_0}{j\omega C_x - j\,\omega^3 LC_0 C_x + j\,\omega\, C_0 C_x/C - \omega^2 RC_0 C_x}$$

$$= \frac{1 - \omega^2 LC_0 + C_0/C + C_x/C - \omega^2 LC_x + j\omega RC_t}{-\omega^2 RC_0 C_x + j(\omega C_x - \omega^3 LC_0 C_x + \omega\, C_0 C_x/C)}$$

The foregoing equation is rationalized by multiplying the denominator and numerator by the complex conjugate of the denominator. The result is

$$Z_{AB} = \frac{N}{\omega^4 R^2 C_0^2 C_x^2 + (\omega C_x - \omega^3 LC_0 C_x + \omega C_0 C_x/C)^2} \tag{72}$$

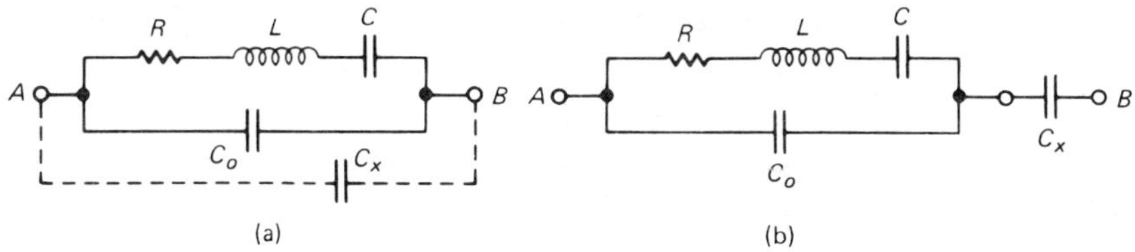

Fig. 7.2. **Equivalent circuit of the crystal unit with load capacitance C_x in (*a*) parallel and (*b*) series.**

where the numerator N is

$$N = \omega^2 RC_x^2 - j\left(\omega^5 L^2 C_0^2 C_x + \omega^5 L^2 C_0 C_x^2 - \frac{2\,\omega^3 LC_0 C_x^2}{C}\right.$$

$$- \frac{2\,\omega^3 LC_0^2\, C_x}{C} - \omega^3\, LC_x^2 + \omega^3 R^2\, C_0 C_x^2 + \omega^3\, R^2 C_0^2\, C_x$$

$$+ \frac{2\,\omega\, C_0 C_x}{C} + \omega\, C_x$$

$$\left. - 2\omega^3 LC_0 C_x + \frac{\omega C_x^2}{C} + \frac{\omega C_0^2 C_x}{C^2} + \frac{\omega\, C_0 C_x^2}{C^2}\right)$$

In order to determine the frequencies of zero phase we set the imaginary part of N equal to zero. The terms in the imaginary part of N which contain R are some six orders of magnitude smaller than any of the other terms and therefore may be neglected in calculating the resonant frequencies. This is in agreement with the well-known fact that the frequency is affected very little by damping in a circuit in which Q is large.

One solution is $\omega = 0$, corresponding to the direct-current case. Dividing through by ω after dropping the terms containing R leaves

$$\omega^4 \left(L^2 C_0^2 C_x + L^2 C_0 C_x^2\right) - \omega^2 \left(\frac{2LC_0 C_x^2}{C} + \frac{2LC_0^2 C_x}{C} + 2LC_0 C_x + \right.$$

$$\left. LC_x^2\right) + \left(\frac{2\, C_0 C_x}{C} + C_x + \frac{C_x^2}{C} + \frac{C_0^2\, C_x}{C^2} + \frac{C_0 C_x^2}{C^2}\right) = 0$$

This expression can be simplified by multiplying through by (C^2/C_x) and letting $C_t = (C_0 + C_x)$ to give (assuming that $C_x \neq 0$)

$$\omega^4 L^2 C^2 C_0 C_t - \omega^2\, [2LCC_0 C_t + LC^2\, (C_0 + C_t)] + [(C_0 + C)(C_t + C)] = 0$$

This equation can be solved for ω^2 by the use of the quadratic formula. The result is

$$\omega^2 = \frac{LC^2(C_0 + C_t) + 2LCC_0C_t \pm LC^2(C_0 - C_t)}{2L^2C^2C_0C_t}$$

Using the + sign,

$$\omega^+ = \sqrt{\frac{1}{LC} + \frac{1}{LC_0}} \tag{73}$$

Using the – sign,

$$\omega^- = \sqrt{\frac{1}{LC} + \frac{1}{LC_t}} \tag{74}$$

The impedances at the frequencies ω^+ and ω^- are found by using Eq. (72). The term involving R in the denominator of Eq. (72) is negligible compared with the other terms, and of course, the imaginary part of N is equal to zero. Substituting ω^- from Eq. (74) for ω in Eq. (72) gives

$$Z_{AB}^- = \frac{R}{(1 - C_0/C_t)^2} \tag{75}$$

The impedance Z_{AB}^+ at frequency ω^+ is found by substituting from Eq. (73) into Eq. (72). In this case, however, all of the denominator except the term involving R is identically zero leaving only

$$Z_{AB}^+ = \frac{1}{\omega^2 RC_0^2} \tag{76}$$

From Eqs. (75) and (76) it is easily shown that Z_{AB}^+ is very large, typically 10^5 to 10^6 Ω; while Z_{AB}^- is of the same order of R as long as C_x is appreciably greater than C_0. It is logical, therefore, to call ω^+ the frequency of antiresonance and ω^- the frequency of resonance. For convenience of comparison, the results are tabulated below for the two cases where C_x is in parallel and in series with the crystal unit.

Parallel (Fig. 7.2*a*)

$$\omega_R = \sqrt{\frac{1}{LC}}$$

$$\omega_A = \sqrt{\frac{1}{LC} + \frac{1}{LC_t}}$$

$$Z_R = R$$

$$Z_A = \frac{1}{\omega^2 C_t^2 R}$$

Series (Fig. 7.2*b*)

$$\omega'_R = \sqrt{\frac{1}{LC} + \frac{1}{LC_t}}$$

$$\omega'_A = \sqrt{\frac{1}{LC} + \frac{1}{LC_0}}$$

$$Z'_R = \frac{R}{(1 - C_0/C_t)^2}$$

$$Z'_A = \frac{1}{\omega^2 C_0^2 R}$$

Two important results of the foregoing derivation should be noted since they form the basis of the methods most generally used to determine the values of L and C. First, the series-resonant (low-impedance) frequency of the series combination is the same as the parallel-resonant (high-impedance) frequency of the parallel combination. This reduces a difficult measurement of f_A to a simple one. The other important result is that the value of R can also be obtained from the measurement of the impedance of the series combination at the low-impedance resonant frequency. If we let $R_e = Z'_R$ be the impedance of the series combination at the resonant frequency, then

$$R = R_e \left(1 - \frac{C_0}{C_t}\right)^2 = R_e \left(\frac{C_x}{C_t}\right)^2 \tag{77}$$

THE PROBLEM OF OSCILLATOR CORRELATION

When crystal units first began to be used to control the frequencies of oscillators in communication equipments, the crystal units were usually fabricated to give the desired frequency in the actual equipment. Since to transport the actual equipment to the site of the crystal manufacturer was often impractical, common practice was to extract the oscillator from the equipment and supply it to the crystal manufacturer.

Circuits which simulated the oscillator were later supplied. Since these "test sets" were often enclosed in cabinets which were painted black, they became known as *black boxes*. A crystal-unit manufacturer would be supplied with one or more black boxes for each type of crystal unit he produced. It was soon recognized that the frequency developed by a crystal unit when operated in one black box did not equal that developed in another black box, which was supposed to be identical with the first. The problem of correlation became more severe as frequency tolerances were decreased.

Ultimately it became clear that the specification of frequency and activity does not adequately describe a crystal unit. Suppose, for example, that a crystal unit is specified to have a certain frequency when operated in a given oscillator at a parallel-resonant frequency. Due to unavoidable manufacturing tolerances the capacitances presented to the crystal unit will be different in different oscillators of the same type. From Eq. (66*b*) it is seen that the frequency f_A can be obtained in an infinite number of ways by varying L, C, and C_0 even if C_x is constant. Suppose now that another crystal unit, specified to have the same frequency, has different values of L, C, and C_0 but produces the same frequency, in the original oscillator. Now if both crystal units are operated in another oscillator having a slightly different C_x the frequencies of the two crystal units will not only differ from the original frequency but may differ from each other.

The solution is to specify not only the frequency but also C_0, and either L or C (but not both). Specified in this way, crystal units are interchangeable among oscillators of the same type with acceptable frequency differences.

The CI-meter was devised to meet the need for an instrument to measure the motional parameters L, C, and R of the crystal unit. A recent modification of the CI-meter has also included a circuit for measuring C_0.

THE CI-METER

The crystal impedance meter, or CI-meter, is similar in principle to the simple circuit of Fig. 7.1. A simplified diagram is shown in Fig. 7.3. It differs primarily from the simpler circuit in that the inductance is divided and the feedback path is provided by a connection

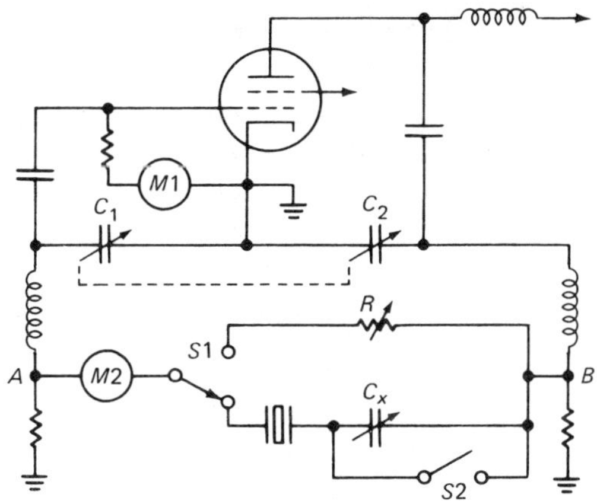

Fig. 7.3. Simplified circuit of the CI-meter.

between points A and B. The circuit is shielded so that oscillation cannot occur except by feedback through the path AB. Switches S_1 and S_2 are provided for convenience and resistor R is a set of calibrated, noninductive decade resistors which may be substituted for the crystal unit.

Meter M_1 reads the grid current of the oscillator and is used in determining the value of R and R_e. When the crystal unit and the resistor result in the same grid current (which is a measure of the amplitude of the oscillation), then the impedance of the crystal unit is equal to that of the resistor. M_2 is a radio-frequency milliammeter which is used to measure the current through the crystal unit. C_x is a calibrated variable capacitor used to supply the load capacitance specified for the particular crystal unit.

The series-resonant frequency and the series resistance are measured with S_2 closed. When the same frequency and grid current are obtained with a resistor and the crystal unit, the frequency is f_R and the resistance R is read from the decade resistor.

To measure the parallel-resonant frequency switch, S_2 is opened and C_x is adjusted to the value specified for the crystal unit. Although C_x is in series with the crystal unit, the frequency of oscillation is the same as the frequency which the crystal unit would produce in an oscillator circuit in which C_x is in parallel with the crystal unit, as we have shown in the previous section. The tuning capacitors C_1 and C_2

and the decade resistor are adjusted until the frequency and grid current are not changed by switching S_1 from the crystal unit to the resistor. When the grid current and the frequency are unchanged, the frequency is f_A and the resistance is R_e. The equivalent resistance R_e may be used as a measure of the quality of the crystal unit. If the value of R is required, it may be calculated using Eq. (77). The impedance which the crystal unit will have in parallel with the load capacitance C_x in the parallel-resonant mode may be found by combining Eqs. (68*b*) and (77). It is

$$Z_A = \frac{1}{\omega^2 C_x^2 R_e} \tag{78}$$

Whereas the CI-meter is simple in theory and convenient to use, requiring only a frequency counter in addition, it does have certain disadvantages. The construction of the instrument is critical because of the necessity of eliminating or compensating all stray reactances. These may cause the readings obtained on different instruments to differ slightly. Nevertheless, since its invention, most crystal units have been specified in terms of measurements made with the CI-meter.

More recently the development of highly stable yet continuously variable sources of radio-frequency signals, called *frequency synthesizers*, and an instrument called the *vector voltmeter*, have made possible measurements on the crystal unit as a passive element. This method has begun to replace the CI-meter, especially for measurements requiring the highest precision and reproducibility.

PASSIVE-NETWORK METHODS

Figure 7.4 shows, in block-diagram form, the system used for measuring the parameters of the equivalent circuit by using the crystal unit as an element in a passive network.

The crystal unit is part of a π network consisting of the low-resistance, noninductive resistors R_1 and R_2. Careful shielding is required to ensure that no current is induced in R_2 by inductive or capacitive coupling with R_1. The inputs to the vector voltmeter are connected to points A and B. The vector voltmeter may be used to

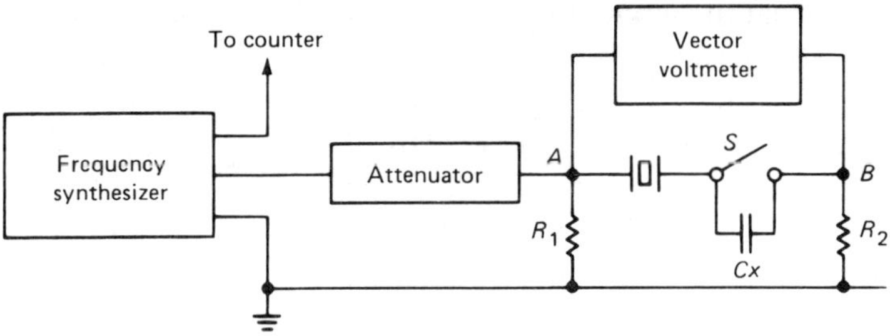

Fig. 7.4. Apparatus used in measuring the parameters of the equivalent circuit by a passive method.

measure the voltage at A or B, but its important function is to measure the phase difference between the voltages at A and B, which is, of course, the phase angle of the impedance between points A and B. It is capable of measuring this phase angle to a fraction of a degree.

We now show that the phase difference between points A and B is an exceedingly sensitive indicator of the resonant frequency of the crystal unit. At frequencies very near f_R the crystal unit appears to be a series-resonant network consisting of R, L, and C in series having impedance

$$Z = R + j\left(\omega L - \frac{1}{\omega C}\right)$$

This impedance can be represented graphically by the triangle shown in Fig. 7.5 in which $\tan\theta = (\omega L - 1/\omega C)/R$. But since the phase angle θ is very small (not over a few degrees), we may replace the tangent of the angle by the angle and write

$$\theta = \frac{\omega L - 1/\omega C}{R}$$

Fig. 7.5. Impedance of the crystal unit very near the resonant frequency f_R.

Differentiating the equation above with respect to ω we have

$$\begin{aligned} d\theta &= \left(L + \frac{1}{\omega^2 C}\right) \frac{d\omega}{R} \\ &= \frac{\omega^2 LC + 1}{\omega^2 RC}\, d\omega = \frac{2}{\omega^2 RC}\, d\omega \\ &= \frac{2\,Q}{\omega}\, d\omega \qquad \text{since } Q = \frac{1}{\omega RC} \end{aligned}$$

If $d\theta$ is expressed in degrees we have

$$d\theta = \frac{360\,Q}{\pi\ \ f}\, df \tag{79}$$

For example, if a 10-MHz crystal unit has a Q of 10^5, a frequency change of 1 Hz causes a phase change of slightly more than one degree, which is well within the sensitivity of the vector voltmeter. If Q is larger, the change in phase angle is proportionally greater also.

In practice the instrument is first set up with a nonreactive resistor in place of the crystal unit. The frequency synthesizer is set to the approximate frequency of the crystal unit. A phase-shifting network in the vector voltmeter is used to compensate for phase shifts in the cables and the π network. When the phase shifter has been set to show zero phase, the crystal unit is placed in the socket and the frequency adjusted to produce zero phase. By definition, this frequency is f_R. The value of R can be determined by substituting for the crystal unit, a resistor which gives the same voltage at point B.

The antiresonant frequency f_A and the equivalent resistance R_e are determined by opening switch S, thereby placing load capacitance C_x in series with the crystal unit. The frequency is adjusted for zero phase, giving the frequency f_A. R_e can be determined by substituting a resistor between points A and B which yields the same voltage at point B as the crystal unit in series with C_x.

Having determined R, f_R, f_A, and C_0, the values of L, C, Q, and Z_A can be determined using the same equations as those used with the CI-meter.

THE EFFECT OF SHUNT CAPACITANCE

When a crystal unit is used as the frequency-controlling element in a Pierce or Miller oscillator circuit it operates at the parallel-resonant frequency f_A, which is given by

$$f_A = \frac{1}{2\pi}\sqrt{\frac{1}{LC} + \frac{1}{LC_t}} \qquad C_t = C_0 + C_x$$

Since the frequency depends on C_t, it follows that any variation in C_t due to a change in C_x must influence the frequency of oscillation. Sometimes the value of C_x is deliberately changed to vary the frequency of the oscillator. The frequency of a typical 5-MHz AT unit can be "warped" or "trimmed" by a few kilohertz in this way.

Usually, however, the variations of frequency which occur as a result of changes in C_t are undesirable and are to be minimized as much as possible.

Differentiating the equation above with respect to C_t gives

$$df_A = -\frac{1}{4\pi LC_t^2\,(1/LC + 1/LC_t)^{1/2}}\, dC_t$$

But since C_t is very large compared with C, we may write

$$df_A = -\frac{\sqrt{LC}}{4\pi LC_t^2}\, dC_t = -\frac{fC}{2C_t^2}\, dC_t \qquad (80)$$

Equation (80) shows that an increase in C_t causes the frequency to decrease. It also shows that the change in frequency is proportional to C, the motional capacitance of the crystal unit. In order to minimize the change of frequency, C should therefore be made small. Such a crystal is often called a *stiff* crystal. The change of frequency with shunt capacitance varies inversely with the square of C_t; and to minimize the frequency change, C_t should be made as large as possible.

Because of variations in components and wiring tolerances, no two oscillator circuits are identical, so a given crystal unit may operate at slightly different frequencies in different oscillators of the same type. Furthermore, aging of components and temperature effects result in changes of the shunt (or load) capacitance with accompanying changes of frequency. These effects are minimized, but not eliminated, by careful selection of components and care in circuit design.

A particularly severe problem is that of changes in the effective value of the shunting capacitance due to dynamic effects. Active elements such as vacuum tubes and transistors are not strictly unilateral devices. Variations in the reactance of the load at the output terminals of tubes and transistors produce variations in the reactance at the input terminals. This phenomenon is often called the Miller effect. The changes of reactance result in changes of C_t and changes in the antiresonant frequency. The effect of dynamic changes of C_t are minimized by keeping the coupling between the crystal unit and the active element as loose as possible. For maximum stability, the crystal unit should be operated at minimum amplitude of oscillation, i.e., at minimum drive level.

LIMITING CONDITION FOR OSCILLATION

The advantage of using a large value of C_t has been discussed in the previous paragraph. However, there is a maximum value of C_t beyond which the crystal unit cannot be used in a circuit which requires that the crystal unit develop an inductive reactance.

We can readily determine the limiting value of C_t by returning to Eq. (63), which may be written (substituting C_t for C_0) as

$$\omega^4 L^2 C^2 C_t - \omega^2 (LC^2 + 2LCC_t - R^2 C^2 C_t) + (C + C_t) = 0$$

The condition for the existence of real roots of the equation is that the discriminant $(b^2 - 4ac) \geqslant 0$. This requires that

$$(4LC^3R^2 - R^4C^4)C_t^2 + 2LC^4R^2C_t - L^2C^4 \leqslant 0 \qquad (81)$$

The inequality above has the form $ax^2 + bx - c \leqslant 0$, which is true only if the value of x lies within the closed interval bounded by the roots of the equation $ax^2 + bx - c = 0$, where a, b, and $c > 0$.

The roots of the equation formed by setting Eq. (81) equal to zero are

$$\frac{LCR \pm 2L\sqrt{LC}}{R^3C - 4LR}$$

But since $LCR \ll 2L\sqrt{LC}$ and $R^3C \ll 4LR$ we may write that the roots are

$$\pm\frac{\sqrt{LC}}{2R}$$

and therefore the condition that Eq. (63) have real roots (and therefore real frequencies of oscillation) is that

$$C_t \leqslant \frac{\sqrt{LC}}{2R} = \frac{1}{2\omega R} = \frac{QC}{2}$$

or, in other words, the maximum value of C_t, which we shall denote C_t^*, is

$$C_t^* = Q\frac{C}{2} \tag{82}$$

If $Q = 10^5$ and $C = 1 \times 10^{-14}$ F, the maximum possible load capacitance is therefore 500 pF. It is a necessary condition that the $C_t \leqslant C_t^*$ but not a sufficient condition for oscillation to occur. The amplifier must be capable of providing a loop gain of 1, that is, it must replace the power dissipated. The circuit may fail to oscillate for either reason.

THE IMPORTANCE OF Q

Equation (82) may be written in another form to say

$$Q^* = 2\frac{C_t}{C}$$

where Q^* is the minimum value Q may have if the crystal is to oscillate with load capacitance C_t. Equation (80) may be written in a different form as follows:

$$\frac{df_A}{f_A} = -\frac{1}{2}\frac{C}{C_t}\frac{dC_t}{C_t} \tag{83}$$

which says that the fractional change in the frequency of oscillation is directly proportional to the fractional change in C_t. The highest stability is achieved when C_t is a maximum. But the maximum value of C_t is set by Eq. (82). Hence, the highest possible frequency stability with respect to variations in C_t is

$$\frac{df_A}{f_A} = \frac{1}{Q}\frac{dC_t}{C_t^*} \tag{84}$$

Equation (84) shows the importance of Q in determining the frequency stability of a crystal-controlled oscillator. Since the Q of an ordinary quartz crystal unit is usually between 10^5 and 10^6, it follows that a 1 percent change in C_t^* results in a frequency change of only 0.1 to 0.01 parts per million. This presupposes that C_t is the maximum possible value, but the actual value will likely be smaller than this by a factor less than 10. The frequency change is, of course, proportionally larger.

THE CRITICAL FREQUENCIES OF THE QUARTZ RESONATOR

It has been shown in Chap. 6 that the quartz crystal unit can be represented by the electric network of Fig. 6.5. We now show that this network has six critical frequencies; each is of importance in the piezoelectric resonator. These six frequencies are:

1. The series-resonant frequency: $f_R = 1/2\pi\sqrt{1/LC}$, [Eq. (65)].
2. The parallel-resonant frequency: $f_A = 1/2\pi\sqrt{1/LC + 1/LC_0}$, [Eq. (66)].
3. The frequency of zero phase f_r, at which Z_{AB} is a pure resistance having a low value.
4. The frequency of zero phase f_a, at which Z_{AB} is a pure resistance having a large value.
5. The frequency at which Z_{AB} is a minimum, f_m.
6. The frequency at which Z_{AB} is a maximum, f_n.

THE FREQUENCIES OF ZERO PHASE

We have previously neglected the effect of dissipation of energy in determining the frequencies of resonance [see Eqs. (63) and (64)]. However the frequencies at which Z_{AB} is a pure resistance do depend to a slight degree on the value of R.

These frequencies can most easily be determined by calculating the admittance of the network of Fig. 6.5 and then setting the imaginary part equal to zero and solving for the frequencies.

Let $Y_1 = j\omega C_0$ be the admittance of the capacitative branch of the equivalent network and let $Y_2 = 1/(R + jX)$ where $X = R + j(\omega L - 1/\omega C)$ is the admittance of the series branch. Then

$$Y_{AB} = Y_1 + Y_2 = \frac{R}{R^2 + X^2} - j\left(\frac{X}{R^2 + X^2} - \omega C_0\right)$$

Setting the imaginary part equal to zero gives $\omega C_0 X^2 - X + \omega C_0 R^2 = 0$. By use of the quadratic formula we find

$$X = \frac{1 \pm \sqrt{1 - 4(\omega C_0 R)^2}}{2\,\omega C_0}$$

Now since $(2\,\omega C_0 R)^2 \ll 1$ we may write

$$X = \frac{1 \pm [1 - 2(\omega C_0 R)^2]}{2\,\omega C_0}$$

Using, first, the minus sign, we have

$$X = \omega C_0 R^2$$

However, $X = \omega L - 1/\omega C$, which can be written

$$X = \frac{\omega^2 LC - 1}{\omega C}$$

Using $\omega_R^2 LC = 1$ we may write

$$X = \frac{\omega^2 LC - \omega_R^2 LC}{\omega C} = \frac{(\omega - \omega_R)(\omega + \omega_R) L}{\omega}$$

But since $(\omega + \omega_R) \doteq 2\omega$, we may write

$$X = 2(\omega - \omega_R) L$$

Equating the two values for X, we have $\omega C_0 R^2 = 2(\omega - \omega_R) L$, from which $\omega_R = \omega(1 - C_0 R^2/2L)$. Then, because $C_0 R^2/2L \ll 1$, we may write $\omega \doteq \omega_R (1 + C_0 R^2/2L)$. This is the lower frequency at which Z_{AB} is a pure resistance. Calling this frequency f_r we have

$$\boxed{f_r = f_R \left(1 + \frac{r}{2Q^2}\right)}$$

where we have written $r = C_0/C$ and $Q = \omega L/R = 1/\omega RC$.

Taking the plus sign, we have

$$X = \frac{2 - 2(\omega C_0 R)^2}{2\omega C_0} = \frac{1 - (\omega C_0 R)^2}{\omega C_0}$$

Just as in the previous case, we set the two values of X equal and solve for ωR. In this way we obtain

$$\omega_R = \omega \left(1 - \frac{1}{2\omega^2 LC_0} + \frac{C_0 R^2}{2L}\right)$$

Now $1 = \omega_R^2 LC$, so

$$\frac{\omega_R^2 LC}{2\omega^2 LC_0} \doteq \frac{C}{2C_0} = \frac{1}{2r}$$

Since the last two terms are each small compared to 1, we may write

$$\omega_a = \omega_R \left(1 + \frac{1}{2r} - \frac{r}{2Q^2}\right)$$

or

$$\boxed{f_a = f_R \left(1 + \frac{1}{2r} - \frac{r}{2Q^2}\right)}$$

The frequency f_a is the antiresonant frequency (or the frequency of parallel resonance including the effect of damping). Comparing with Eq. (66)[1] in which the effect of damping is neglected, we see that the effect of damping is to make $f_a < f_A$. (We have just seen that the effect of damping is to make $f_r > f_R$.) Hence, the effect of damping is to reduce the frequency difference between f_a and f_r. The effect is very small, since it depends inversely on Q^2. Typically f_A and f_a, or f_R and f_r, differ by a few parts in 10^8 in high-frequency fundamental-mode AT-cut units.

FREQUENCIES OF MINIMUM AND MAXIMUM IMPEDANCE

In order to determine the frequencies at which the impedance Z_{AB} has a minimum and a maximum value, we first compute the modulus of the admittance, which is

$$\left|Y_{AB}\right|^2 = \left(\frac{R}{R^2+X^2}\right)^2 + \left(\frac{X}{R^2+X^2} - \omega C_0\right)^2$$

$$= \frac{1 - 2\,\omega C_0 X}{R^2+X^2} + \omega^2 C_0^2$$

In order to find the minimum and maximum values of $|Y_{AB}|^2$, we differentiate the expression with respect to X (which is a function of ω) and set the result equal to zero. The result is

$$\omega C_0 X^2 - X - \omega C_0 R^2 = 0$$

Using the quadratic formula to find the values of X gives

$$X = \frac{1 \pm \sqrt{1 + (2\,\omega C_0 R)^2}}{2\,\omega C_0} \approx \frac{1 \pm [1 + 2\,(\omega C_0 R)^2]}{2\omega C_0}$$

---

[1] Equation (66) may be written

$$\omega_A^2 = \frac{1}{LC} + \frac{1}{LC_0} = \frac{1}{LC}\left(1 + \frac{1}{r}\right) = \omega_R^2\left(1 + \frac{1}{r}\right)$$

or

$$\omega_A \doteq \omega_R\left(1 + \frac{1}{2r}\right) \quad \text{and} \quad f_A \doteq f_R\left(1 + \frac{1}{2r}\right)$$

Using first the minus sign, we have $X = -\omega C_0 R^2$.
But also $X = 2(\omega - \omega_R)L$, so we have, by setting equal the two expressions for X,

$$\omega = \omega_R \frac{2L}{2L + C_0 R^2} \doteq \omega_R \left(1 - \frac{C_0 R^2}{2L}\right)$$

This is the frequency of minimum impedance ω_m. Therefore, $f_m = f_R\ (1 - C_0 R^2/2L)$, or

$$\boxed{f_m = f_R \left(1 - \frac{r}{2Q^2}\right)}$$

Now using the plus sign we find

$$X = \frac{1 + (\omega C_0 R)^2}{\omega C_0}$$

Again setting this equal to $X = 2(\omega - \omega_R)L$ we have that

$$\omega_R = \omega \left(1 - \frac{1}{2\,\omega^2 L C_0} - \frac{C_0 R^2}{2L}\right)$$

Since the last two terms are very small compared with 1, we may write with very small error that

$$\omega \doteq \omega_R \left(1 + \frac{1}{2\,\omega^2 L C_0} + \frac{C_0 R^2}{2L}\right)$$

This is the frequency of maximum impedance, so we may write

$$\boxed{f_n = f_R \left(1 + \frac{1}{2r} + \frac{r}{2Q^2}\right)}$$

All of these results may be briefly summarized in the following way:

1. The series-resonant frequency: $f_R = 1/2\pi\ \sqrt{1/LC}$.
2. The parallel-resonant frequency: $f_A = 1/2\pi \sqrt{1/LC + 1/LC_0}$ $\doteq f_R\ (1 + 1/2r)$.
3. The lower frequency of zero phase: $f_r = f_R\ (1 + \Delta)$.
4. The upper frequency of zero phase: $f_a = f_R\ (1 + 1/2r - \Delta)$.
5. The frequency of minimum impedance: $f_m = f_R\ (1 - \Delta)$.

6. The frequency of maximum impedance: $f_n = f_R\ (1 + 1/2r + \Delta)$, where $\Delta = r/2Q^2$ and $r = C_0/C$ (or C_t/C, where $C_t = C_0 + C_x$). [See Eq. (66*b*).]

The results may also be shown graphically as follows.

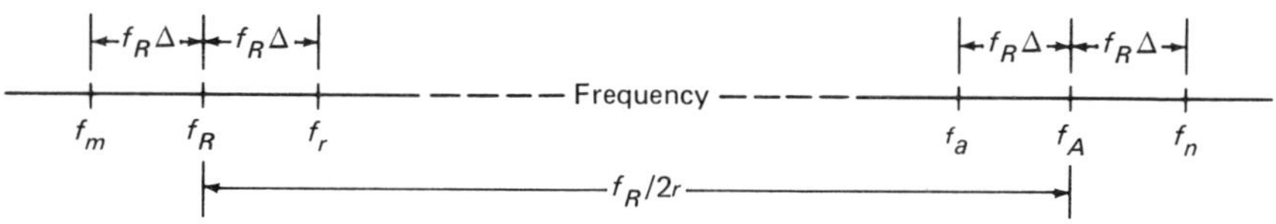

THE IMPEDANCE DIAGRAM

At the angular frequency $\omega = 2\pi f$ the impedance Z of the circuit consisting of an inductance L, a capacitance C, and a resistance R in series is given by

$$Z = R + jX$$

where $X = (\omega L - 1/\omega C)$
If $\omega L > 1/\omega C$ then X is positive and the circuit has an inductive reactance. Conversely, if X is negative the circuit has a capacitive reactance. If $\omega_R^2 LC = 1$ the angular frequency ω_R is called the series resonant frequency.

The impedance Z can be represented by a vector in the complex plane (Fig. 7.6). The real component of OP, which is OA, represents the resistance R and the imaginary component AP represents the re-

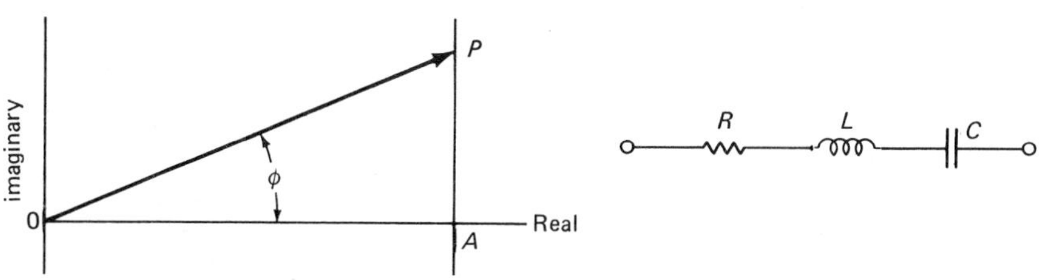

Fig. 7.6. The impedance diagram for the series circuit R, L, C.

actance X. As the frequency is varied from 0 to ∞ the point P moves *up* the line AP from $-\infty$ to $+\infty$.

It is possible to calibrate the line AP for frequency. In the immediate neighborhood of A, corresponding to f_R, the calibration is almost linear. This may be shown in the following way. The reactance

$$\begin{aligned} X &= \omega L - 1/\omega C = (\omega^2 LC - 1)/\omega C = (\omega^2 LC - \omega_R^2 LC)/\omega C \\ &= (\omega - \omega_R)(\omega + \omega_R)(L/\omega) \\ &= 2\,\Delta\omega L \end{aligned}$$

where $\Delta\omega = \omega - \omega_R$ and we have taken $(\omega + \omega_R) = 2\omega$ Therefore

$$X = 4\,\pi L\,\Delta f$$

and the distance AP which is proportional to X is also proportional to $(f - f_R) = \Delta f$.

The phase angle ϕ is given by $\phi = \tan^{-1}\,(AP/OA)$

THE ADMITTANCE DIAGRAM

The admittance Y is the reciprocal of the impedance Z so that

$$\begin{aligned} Y &= 1/Z = 1/(R + J\,X) \\ &= R/(R^2 + X^2) - jX/(R^2 + X^2) \\ &= g - j\,b. \end{aligned}$$

The quantity $g = R/(R^2 + X^2)$ is called the conductance and the quantity $b = X/(R^2 + X^2)$ is called the susceptance. The admittance $Y = g - j\,b$ can also be represented by a vector in the complex plane by making a suitable choice of scale factors. It must be noted, however, that the signs of the imaginary parts of Z and Y are reversed as shown in the following chart.

		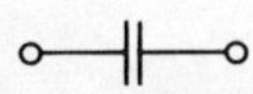
Reactance	+	–
Susceptance	–	+

To construct an admittance diagram for the series circuit L, C, R we let 0 be the origin of coordinates in the complex plane (Figure 7.7). We draw a circle with diameter OA making $R = s_i \cdot OA$ where s_i is some convenient scale factor (ohms/cm). The circle is called the admittance circle. A line is drawn through point A perpendicular to OA. The triangle OPA is the impedance triangle. We then mark point Q where line OP intersects the circle. As the point P moves along the line AP the point Q moves around the circle. Hence to every point on the line AP there corresponds a point on the admittance circle. We shall now show that the distance OQ represents the admittance Y, the line QM the (negative of the) susceptance b, and the line OM the conductance g. To do so we require a theorem from analytic geometry which says that

$$OP \cdot OQ = (OA)^2$$

The proof of this theorem is relatively simple. We draw the line QA. Then in right triangle OQA

$$(OQ)^2 + (QA)^2 = (OA)^2$$

In right triangle OPA

$$(OP)^2 - (PA)^2 = (OA)^2$$

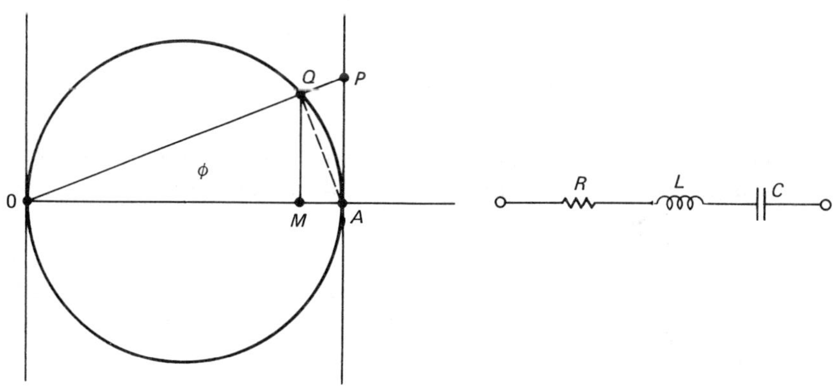

Fig. 7.7. Admittance diagram of the series circuit R, L, C.

and in right triangle *QAP*

$$(PA)^2 - (QA)^2 - (QP)^2 = 0$$

Obviously

$$(QP)^2 = (OP - OQ)^2 = (OP)^2 + (OQ)^2 - 2\, OP \cdot OQ$$

Adding the first three equations and substituting from the last gives the required proof.

In the impedance triangle *OPA* we have $R = s_i \cdot OA$, $X = s_i \cdot AP$ and $Z = s_i \cdot OP$ where s_i is the impedance scale factor (ohms/cm).

In the admittance triangle *OQM* we take $g = s_a \cdot OM$, $-b = s_a \cdot QM$, and $Y = s_a \cdot OQ$ where s_a is the admittance scale factor (mhos/cm). From the above theorem we have

$$(Z/s_i) \cdot (Y/s_a) = (R/s_i)^2$$

from which

$$s_a = s_i/R^2 \qquad \text{(since } Z \cdot Y = 1)$$

To calibrate the line *AP* for frequency we take

$$X = 4\,\pi\, L\, \Delta f = s_f\, \Delta f$$

from which

$$AP = \frac{4\,\pi L}{s_i}\, \Delta f = s_f \Delta f$$

the quantity $s_f = 4\,\pi\, L/s_i$ is called the frequency scale factor (cm/Hz).

If care is taken in drawing the diagram to scale the approximate values of the various quantities can be determined by measuring the lengths of the corresponding lines and multiplying by the appropriate scale factors. It must be remembered that the signs of *X* and *b* are reversed and that the phase angles of *Z* and *Y* are also opposite although they are equal in magnitude.

We may now illustrate the use of the admittance-impedance diagram by considering the series circuit of R = 20 ohms, L = 0.031 *H*, and $C = 33 \times 10^{-14}$ F. The figures are typical values of the motional parameters of a 5-MHz AT-cut quartz crystal unit.

We first choose a convenient impedance scale factor s_i, say 1 ohm/cm, and draw the line OA having a length $R/s_i = 20$ cm. A circle with diameter OA is then drawn and the line perpendicular to OA through the point A is constructed.

Suppose that we desire to know the values of R, X, Z, g, b, Y and ϕ at the frequency 10 Hz above f_R. We first compute the value of the frequency scale factor $s_f = 4\pi L/s_i$ which is found to be 0.39 cm/Hz and mark the point P making $AP = 3.9$ cm. Then by direct measurement and use of the scale factors we find

$$R = s_i \cdot OA = 20 \text{ ohms} \qquad g = s_a \cdot OM = 0.048 \text{ mhos}$$
$$X = s_i \cdot AP = 3.9 \text{ ohms} \qquad -b = s_a \cdot QM = 0.0098 \text{ mhos}$$
$$Z = s_i \cdot OP = 20.4 \text{ ohms} \qquad Y = s_a \cdot OQ = 0.049 \text{ mhos}$$
$$\phi = \tan^{-1} (AP/OP) = 11°$$

It is considerably easier and quicker to determine the values of the various quantities in this way than by using algebraic methods and for many purposes the results are sufficiently accurate. A resonance curve can easily be made by plotting the admittance against the frequency, both being obtained from the diagram of Fig. 7.7.

The principal advantage of the admittance method is found in dealing with circuits having parallel branches such as the equivalent circuit of the piezoelectric resonator. Sometimes it is a little inconvenient to make the drawings for these circuits to scale because the scale factors become too large and too small but even in such cases the diagram serves as a useful guide to the computations and it also provides many interesting and valuable insights into the behavior of the circuit in the neighborhood of the resonant frequency. We now examine the admittance diagram for such a circuit.

ADMITTANCE DIAGRAM OF THE PIEZOELECTRIC RESONATOR

The admittance of the equivalent circuit of the piezoelectric resonator (Fig. 6.5) is the vector sum of the admittances of the two branches. That is,

$$Y_{AB} = Y + Y_0$$

where $Y = g - j\,b$ and $Y_0 = -j\,b_0 = -j\omega C_0$. In case there is a parallel load capacitance C_x then $C_t = C_0 + C_x$ must be used instead of C_0.

To construct the admittance circle diagram of the circuit of Fig. 6.5 we let the point 0 (Fig. 7.8) be the origin of coordinates in the complex plane. A convenient scale factor s_a is chosen making the line $OO' = s_a \cdot \omega\,C_0$ and draw the line OO', representing the susceptance of C_0, in the upward direction. This convention results in making the vectors which represent capacitive susceptances point upward and those representing inductive susceptances to point downward.

The actual length of OO' varies with the frequency but over the narrow range near f_R in which we are interested it may be considered to be constant.

The admittance circle having diameter $O'A' = 1/(s_a R)$ and the frequency line $A'P$ are drawn. The frequency scale factor $s_f = 4\pi\,L/s_a R^2$ (cm/Hz) is calculated and the frequency line is calibrated by making $A'P = s_f(f - f_R)$ with point A' corresponding to the resonant frequency f_R. As the frequency increases the point P moves *downward* along the line AP and the point Q moves clockwise around the admittance circle. At any given frequency the vector sum $(OO' + O'Q) = OQ$

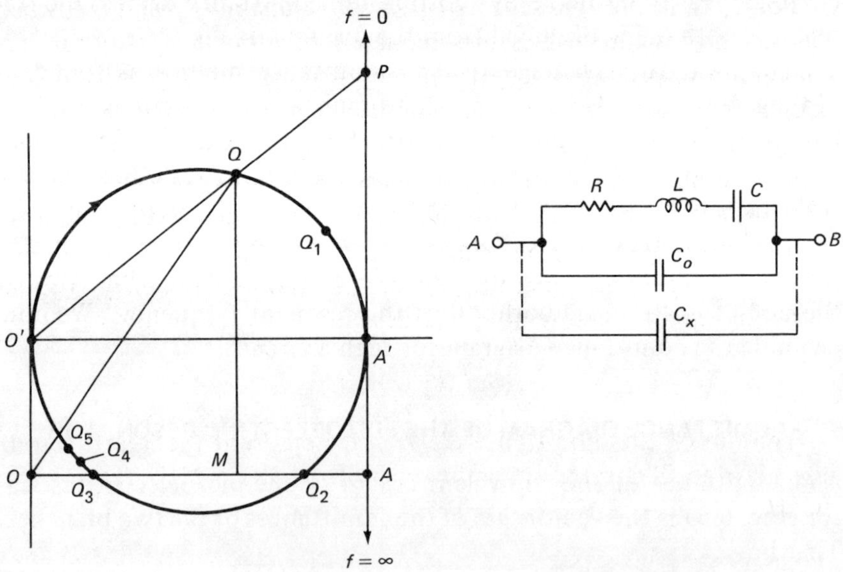

Fig. 7.8. The admittance diagram of the piezoelectric resonator.

represents the admittance of the circuit, OM represents the conductance and QM represents the susceptance. We therefore have

$$Y_{AB} = s_a \, OQ$$
$$g = s_a \, OM$$
$$-b' = -(b + b_0) = s_a \, QM$$
$$\tan \phi = (QM/OM)$$

The point Q_1 on the admittance circle is the point at which the line OQ has the maximum length. The corresponding frequency is therefore the frequency of maximum admittance or minimum impedance. As we have seen before the frequency of minimum impedance is slightly below f_R: The difference between the two frequencies is given by $f_m - f_R = A'P_1/s_f$.

The point Q_2 is the lower frequency of zero phase f_r and the point Q_3 is f_a the upper frequency of zero phase. Both frequencies are higher than f_R since frequencies increase downward on the frequency line and clockwise around the admittance circle. Between f_r and f_a the susceptance is negative which means that the reactance is positive or that the crystal unit develops an inductive reactance.

Point Q_4 is the frequency of parallel resonance f_A and Q_5, where the line OQ is shortest, is the frequency of minimum admittance or maximum impedance. The frequency scale in this region is so compressed that it is usually not practical to make geometric measurements.

The actual frequencies corresponding to points $Q_1, Q_2 \ldots$ can be determined by extending the lines $O'Q_n$ to intersect the frequency line, measuring the distance $A'P_n$ and using $A'P_n = s_f \Delta f$ where $\Delta f = (f - f_R)$. The phase angle at the corresponding frequency is found either by direct measurement or from

$$\tan \phi = QM/OM$$

The use of the admittance diagram to represent the quartz resonator may be illustrated with the example of the 5 MHz resonator in which

$$R = 20 \text{ ohms} \qquad C = 3.3 \times 10^{-14} \text{ F}$$
$$L = 0.031 \text{ H} \qquad C_t = 32 \times 10^{-12} \text{ F}$$

We let the impedance scale factor s_i = 0.4 ohms/cm. This makes the admittance scale factor $s_a = s_i/R^2 = 0.001$ mhos/cm. The length of the line OO' which represents the susceptance of C_t is therefore $\omega\, C_t/s_a = 1.00$ cm.

The diameter of the admittance circle $O'A' = R/s_i$ or $(R \cdot s_a)^{-1}$ = 50 cm.

The frequency scale factor $s_f = 4\pi\, L/s_i = 0.97$ cm/Hz.

We draw the line OO' representing the susceptance of C_t; the line $O'A'$ which is the diameter of the admittance circle; and calibrate the frequency line $A'P$.

To illustrate how the diagram may be used we suppose that it is desired to know the admittance and phase angle of the circuit at a frequency 10 Hz below the frequency f_R. The point P is marked making $A'P$ = 9.7 cm with P above point A'. The line $O'P$ is drawn and the point Q marked on the admittance circle. QM and OQ are drawn to form triangle OQM. Then by direct measurement it is found that OQ = 49.5 cm, OM = 48.4 cm and QM = 10.4 cm from which Y_{AB} = 0.049 mhos, g = 0.048 mhos and b' = 0.010 mhos. The phase angle ϕ measured by use of a protractor or computed from tan $\phi = QM/OM$ is 12°. Since the susceptance is positive the circuit has a capacitive reactance at this frequency.

It may be surprising to find that the phase angle changes so rapidly in the neighborhood of the series resonant frequency f_R but it is this characteristic which enables the crystal unit to control the frequency of an oscillator circuit with such great accuracy.

As the value of C_t is increased the length of the line OO' increases and the points Q_2 and Q_3 move closer together finally meeting at a point where the admittance circle is tangent to the line OA. In this case length of OO' is equal to the radius of the admittance circle and therefore

$$C_t^*/s_a = \frac{1}{2}\,(1/s_a R)$$

or

$$C_t^* = 1/2\,\omega R = QC/2$$

which is the same result obtained by algebraic methods in Eq. (82). C_t^* is the maximum value of C_t for which the crystal unit is able to develop an inductive reactance.

Many other interesting and instructive characteristics of the quartz crystal resonator may be found by considering the admittance circle diagram.

8
THE FREQUENCY CONSTANTS AND TEMPERATURE COEFFICIENTS OF FREQUENCY

INTRODUCTION

In this chapter we show how the frequency constant of a given piezoid is determined by the density, the dimensions and the elastic constants. Then we show how the temperature coefficient of the resonant frequency is determined by the temperature dependence of the same factors.

THE FREQUENCY CONSTANTS

We have seen [Eqs. (39), (40), and (43)] that the fundamental and the harmonic resonant frequencies of an infinite plate, i.e., one in which the length and width, or diameter, are very large compared with the thickness, is given approximately by

$$f_n = \frac{nv}{2e} \qquad n = 1, 3, 5, \ldots$$

where v is the velocity with which the longitudinal or shear wave is propagated in the plate and e is the thickness. These are the "harmonic" frequencies utilized in high-frequency quartz crystal units.

The velocity of propagation of a wave in the thickness direction of a thin plate is given by

$$v = \sqrt{\frac{c_{ij}}{\rho}}$$

where c_{ij} is the ratio of the stress to the strain associated with the elastic wave which is being propagated in the quartz and ρ is the density of quartz.

In the case of an X-cut piezoid in which the wave is a longitudinal wave in the thickness direction, the appropriate elastic modulus is c_{11}. For a Y-cut plate, in which the wave is transverse, the stress and strain are related by the modulus c_{66}. If the plate is cut in such a way that the thickness is not parallel to either the X- or Y-axis, as in the AT- and BT-cut plates, the applicable modulus becomes a function of two or more of the elastic moduli and of the angles between the crystallographic axes and the direction of propagation of the wave. The value of the applicable elastic modulus is then determined by the use of matrix algebra, as outlined in Chap. 2.

The resonant frequency of an infinite plate vibrating in a thickness mode is therefore

$$f_n = \frac{n}{2e}\sqrt{\frac{c_{ij}}{\rho}} \qquad n = 1, 3, 5, \ldots \tag{85a}$$

Since all the quantities on the right-hand side of Eq. (85) are constant except e, we may write

$$f_n = n\left(\frac{K}{e}\right) \tag{85b}$$

where $K = \frac{1}{2}\sqrt{c_{ij}/\rho} = \frac{1}{2}\,v$ is called the *frequency constant.* K has the dimensions of velocity but is usually expressed in kilohertz-inches or kilohertz-millimeters.

The values of K are the same if expressed in megahertz-mils (a mil is 0.001 in) or in megahertz-microns (a micron is 0.001 mm). Thus the frequency of a given plate is found by dividing the thickness into the frequency constant. For example, an AT-cut plate having a thickness of 10.0 mils has a frequency of 65.5/10.0 = 6.55 MHz.

Table 8.1 Frequency Constants of a Few Quartz Piezoids.

	K, kHz·in	K, kHz·mm
X-cut	112.1	2849
Y-cut	76.9	1954
AT-cut	65.5	1664
BT-cut	100.3	2549
SC-cut	70.9	1800

The frequencies of the harmonic modes ($n = 1, 3, 5, \ldots$) are not precisely integral multiples of each other in actual resonators. This is due to the differences between the vibrating areas and the plated areas in the different modes. The harmonic frequency f_n may be either slightly larger or smaller than nf_1.

LINEAR COEFFICIENTS

Many physical quantities are approximately linear functions of other physical quantities. Ohm's law, Hooke's law, and the gas law are examples in which a linear approximation is sufficiently accurate for many purposes. When greater accuracy is required the relationship may be expressed as a power series with sufficient terms to provide the required accuracy.

The temperature dependence of the thermal expansion of a solid can be expressed by the equation

$$L = L_0 \, [1 + \alpha (T - T_0) + \beta (T - T_0)^2 + \cdots]$$

where L_0 is the length at temperature T_0 and L is the length at temperature T. For many purposes we may keep only the first term and write

$$L = L_0 \, [1 + \alpha (T - T_0)]$$

or

$$L - L_0 = L_0 \, \alpha (T - T_0)$$

from which

$$\alpha = \frac{1}{L_0} \, \frac{L - L_0}{T - T_0} = \frac{1}{L_0} \, \frac{\Delta L}{\Delta T}$$

The constant α is called the *temperature coefficient of linear expansion.*

THE TEMPERATURE COEFFICIENT OF FREQUENCY

Since the frequency of a resonator depends on its thickness, density, and the elastic constant and each of these changes with the temperature, it follows that the frequency is also temperature-dependent. To a first approximation we may treat each of the temperature dependences as linear and then compute the linear temperature coefficient of the frequency. To do so we write Eq. (85*a*) in logarithmic form in the following way.

$$\log f_n = \log \left(\frac{n}{2}\right) - \log e + \tfrac{1}{2} \log c_{ij} - \tfrac{1}{2} \log \rho$$

We then differentiate the resulting equation with respect to the temperature to obtain

$$T_f = \frac{1}{f}\frac{df}{dT} = -\frac{1}{e}\frac{de}{dT} + \frac{1}{2}\frac{1}{c_{ij}}\frac{c_{ij}}{dT} - \frac{1}{2}\frac{1}{\rho}\frac{d\rho}{dT} \tag{86}$$

The quantity on the left is called the *temperature coefficient* (actually it is the linear temperature coefficient of the frequency) and it is usually expressed in parts per million (ppm) per degree Celsius, or in percent change in frequency per degree Celsius.

Equation (86) indicates that T_f is independent of the harmonic order n, which is true to a first approximation. However, since actual plates are far from infinite, the value of T_f is slightly dependent on n, so plates of a given size are cut at slightly different angles for operation on different harmonic modes.

Since quartz expands when heated, it is easily seen that $d\rho/dT$, the change of density with temperature, is a negative quantity, while the change in thickness with temperature de/dT is a positive quantity and thus from Eq. (86) their effects on T_f are opposite in effect. While the effects do not cancel, it does suggest that the value of T_f might be made zero by balancing out the effects of temperature on these and the elastic modulus. It is fortunate that this is possible in quartz. The usefulness of several otherwise useful crystals is limited by the impossibility of finding any orientation at which the temperature coefficient is zero.

The temperature coefficients of linear expansion in the directions perpendicular and parallel to the Z-axis are called α_1 and α_3, respectively. We have already shown in Chap. 2 that the coefficient of linear expansion is a direction making an angle θ with the Y-axis is given by

$$\alpha' = T_e = \alpha_1 \cos^2\theta + \alpha_3 \sin^2\theta \tag{87}$$

where $\alpha_1 = 14.3 \times 10^{-6}$ and $\alpha_3 = 7.8 \times 10^{-6}/°C$.

With these two values and Eq. (87) the value of the T_e can be calculated for any direction in quartz.

The value of T_ρ can also be calculated. Consider a small cube of quartz having dimensions x, y, and z in the X-, Y-, and Z-directions, respectively. Let m be the mass of the cube. The density is thus

$$\rho = \frac{m}{xyz}$$

The temperature coefficient of the density T_ρ is found as before by taking the logarithmic derivative to obtain

$$\frac{1}{\rho}\frac{d\rho}{dT} = T_\rho = \frac{1}{m}\frac{dm}{dT} - \frac{1}{x}\frac{dx}{dT} - \frac{1}{y}\frac{dy}{dT} - \frac{1}{z}\frac{dz}{dz}$$

But

$$\frac{dm}{dT} = 0 \qquad \frac{1}{x}\frac{dx}{dT} = \frac{1}{y}\frac{dy}{dT} = \alpha_1 \quad \text{and} \quad \frac{1}{z}\frac{dz}{dT} = \alpha_3$$

Therefore the temperature coefficient of the density is

$$T_\rho = -(2\,\alpha_1 + \alpha_3) = -36.4 \times 10^{-6}/°C \tag{88}$$

The values of the temperature coefficients of the elastic moduli have been determined by Mason.[1] His values are given in Table 8.2. It should be noted that all the values are negative except Tc_{66}. It

[1] W. P. Mason, *Piezoelectric Crystals and Ultrasonics.* New York: Van Nostrand, 1950, p. 103.

Table 8.2. Values of the Temperature Coefficients of the Elastic Constants of Quartz.

$Tc_{11} = -46.5 \times 10^{-6}\ °C^{-1}$	$Tc_{33} = -205$
$Tc_{12} = -3300$	$Tc_{44} = -166$
$Tc_{13} = -700$	$Tc_{66} = +164$
$Tc_{14} = -90$	

is due to the fact that this temperature coefficient is positive that a zero temperature coefficient of frequency is possible in quartz.

We now have all the information necessary to compute the value of T_f for the X- and Y-cut plates. The thickness of the X-cut plate is in the X-direction and therefore $T_e = \alpha_1 = 14.3 \times 10^{-6}$. The elastic modulus is c_{11} and $Tc_{11} = -46.5 \times 10^{-6}$. Therefore, for the X-cut plate,

$$T_f = \left[\tfrac{1}{2}(-46.5) - 14.3 - \tfrac{1}{2}(-36.4)\right] \times 10^{-6} = -19.4 \times 10^{-6}\ °C^{-1}$$

which is in agreement with the measured value.

For the Y-cut plate we have that the thickness is in the Y-direction and therefore $T_e = \alpha_1 = 14.3 \times 10^{-6}$. The elastic modulus is c_{66} and Tc_{66} is $+ 164 \times 10^{-6}$. Therefore, for the Y-cut plate, vibrating in thickness shear,

$$T_f = \left[\tfrac{1}{2}(164) - (14.3) - \tfrac{1}{2}(-36.4)\right] \times 10^{-6} = 86 \times 10^{-6}\ °C^{-1}$$

To determine T_f for a plate in which the direction of wave propagation is not parallel to one of the crystallographic axes, it is necessary to determine the temperature coefficient of expansion in the thickness direction of the plate and the temperature coefficient of the applicable elastic modulus. The temperature coefficient of the density is, of course, independent of direction.

We illustrate the process by computing the T_f for a rotated Y-cut plate, i.e., an orientation reached by considering a Y-cut plate to be rotated about the X-axis which lies in the plane of the plate.

The angle of rotation θ is measured as shown in Fig. 8.1a. This angle convention conforms to the convention used by Bond[2] and to the usual method of specifying the orientation of AT, BT, and other

[2] *B. S. T. J.*, 22. (1):12 (January 1943).

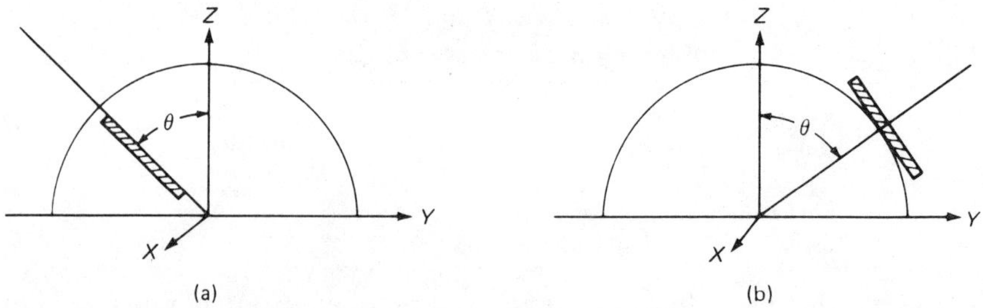

Fig. 8.1. Methods of specifying the angular rotation of piezoids of the *Y*-cut family. (*a*) System used by Bond. (*b*) IRE (IEEE) system.

piezoids. It differs from the convention adopted by the Technical Committee on Piezoelectric Crystals in the 1945 report of the Institute of Radio Engineers (now IEEE). The latter convention is illustrated in Fig. 8.1*b*. Although the IEEE convention is preferable in some respects, the older system is used to avoid the necessity of converting all the rotation matrices which have been given by Bond[3]. Calculations are most easily made in the older system and the results may be expressed according to the IEEE convention where desirable.

The calculation of the values of the elastic moduli for rotated plates is somewhat more complicated than that for the dielectric constants because the elastic moduli are described by 6 X 6 matrices while the dielectric constants are described by 3 X 3 matrices. However the procedure is the same in principle.

The stress matrix transforms according to the relation

$$X' = \alpha X$$

where α is the matrix which performs a rotation about the X axis.

The strain matrix transforms as

$$x' = \alpha_c^{-1} x \qquad \text{or} \qquad x = \alpha_c x'$$

[3] (*Loc. cit.*)

where α_c^{-1} is the reciprocal matrix of the congugate matrix of α. All the required matrices have been tabulated by Bond.[4] Using the methods of matrix algebra developed in Chap. 2 we have

$$X = c\,x$$
$$X' = \alpha c\,x$$
$$X' = \alpha\, c\, \alpha_c\, x'$$

or

$$X' = c'\,x'$$

where c' is written for $\alpha\, c\, \alpha_c$.

The matrix c' corresponding to a rotation by an angle θ about the X axis is found by performing the following matrix multiplications.

$$\begin{pmatrix} 1 & 0 & 0 & 0 & 0 & 0 \\ 0 & c^2 & s^2 & 2sc & 0 & 0 \\ 0 & s^2 & c^2 & -2sc & 0 & 0 \\ 0 & -sc & sc & c^2-s^2 & 0 & 0 \\ 0 & 0 & 0 & 0 & c & -s \\ 0 & 0 & 0 & 0 & s & c \end{pmatrix} \begin{pmatrix} c_{11} & c_{12} & c_{13} & c_{14} & 0 & 0 \\ c_{12} & c_{11} & c_{13} & -c_{14} & 0 & 0 \\ c_{13} & c_{13} & c_{33} & 0 & 0 & 0 \\ c_{14} & -c_{14} & 0 & c_{44} & 0 & 0 \\ 0 & 0 & 0 & 0 & c_{44} & c_{14} \\ 0 & 0 & 0 & 0 & c_{14} & c_{66} \end{pmatrix} \begin{pmatrix} 1 & 0 & 0 & 0 & 0 & 0 \\ 0 & c^2 & s^2 & -sc & 0 & 0 \\ 0 & s^2 & c^2 & sc & 0 & 0 \\ 0 & 2sc & -2sc & c^2-s^2 & 0 & 0 \\ 0 & 0 & 0 & 0 & c & s \\ 0 & 0 & 0 & 0 & -s & c \end{pmatrix}$$

where, as usual, c stands for $\cos\theta$ and s for $\sin\theta$.

Fortunately, to determine a single element of a matrix, it is not necessary to compute the entire matrix. In this case we are interested only in c'_{66}, which is the coefficient required to determine the frequency constant and the temperature coefficients of all members of the rotated Y-cut family of piezoids.

To obtain the value of c'_{66} we need compute only the last row of the matrix obtained by multiplying α by c. This row is found to be

$$0 \quad 0 \quad 0 \quad 0 \quad (c_{44}s + c_{14}c) \quad (c_{14}s + c_{66}c)$$

---

[4] (*Loc. cit.*)

We then multiply the row just obtained by the last column in α_c. This gives

$$c'_{66} = c_{44}s^2 + 2\,c_{14}sc + c_{66}c^2 \tag{89}$$

For $\theta = 0$ corresponding to a Y-cut, $c'_{66} = c_{66}$; and if $\theta = 90°$ corresponding to a Z-cut, $c'_{66} = c_{44}$. Both are correct.

The value of c'_{66} for any angle of rotation about the X-axis can be calculated by using Eq. (89) and the values of the elastic moduli from Table 3.1. In this way the value of c'_{66} for the AT-cut ($\theta = 35\frac{1}{4}°$) is found to be

$$c'_{66} = 29.3 \times 10^9 \text{ N/m}^2$$

and for the BT-cut ($\theta = -49°$)

$$c'_{66} = 68.9 \times 10^9 \text{ N/m}^2$$

We are now able to compute the frequency constants using Eq. (85b) and the density of quartz, which is 2650 kg/m^3.

It must be remembered, however, that the elastic constants are themselves temperature-dependent and therefore c'_{66} is also temperature-dependent. The temperature coefficient of c'_{66} is calculated in the following way. We differentiate Eq. (89) with respect to T and divide both sides by c'_{66}. This gives

$$Tc'_{66} = \frac{1}{c'_{66}}\,\frac{dc'_{66}}{dT} = \frac{s^2\,c_{44}\,Tc_{44} + c^2\,c_{66}\,Tc_{66} + 2sc\,c_{14}\,Tc_{14}}{c'_{66}} \tag{90}$$

The values of the temperature coefficients of the elastic constants c_{66}, c_{44}, and c_{14} have been determined experimentally and are given in Table 8.2. From Eq. (86) we have for the rotated Y-cut family of quartz plates:

$$T_f = -T_e + \tfrac{1}{2}\,Tc'_{66} - \tfrac{1}{2}\,T_\rho \tag{91}$$

Substituting from Eqs. (87), (88), and (90) into Eq. (91) gives the temperature-frequency coefficient for the entire Y-cut family.

$$T_f = -(\alpha_1 c^2 + \alpha_3 s^2) + \tfrac{1}{2}\,\frac{s^2 c_{44} Tc_{44} + c^2 c_{66} Tc_{66} + 2sc\, c_{14} Tc_{14}}{c_{44} s^2 + c_{66} c^2 + 2sc\, c_{14}} + \tfrac{1}{2}(36.4 \times 10^{-6}) \qquad (92)$$

The values of the angle θ for which the temperature-frequency coefficient is zero is found, in principle, by setting the equation above equal to zero and solving for θ. It would be a very laborious task to solve this equation in closed form. With the aid of a computer the values can be obtained readily or solutions can be obtained by successive approximation methods.

If $\theta = 0$ is substituted into Eq. (92), we find $T_f = 86 \times 10^{-6}\ ^\circ\mathrm{C}^{-1}$. This is, of course, the temperature coefficient of the Y-cut piezoid.

If $\theta = 90^\circ$, corresponding to the Z-cut, is substituted, the value of $T_f = -73 \times 10^{-6}$. It is not possible to excite the Z-cut plate into oscillation through the piezoelectric effect because no piezoelectric strains are produced in quartz by electric fields in the Z-direction. However, if the thickness shear mode were excited by driving the plate in some way, it would have a negative temperature coefficient of about −73 ppm/°C. Since the temperature coefficient changes from positive to negative as θ is changed from 0° to 90°, it follows that it must be zero at some point between.

The value of T_f actually becomes zero at two values of θ. These are at approximately $\theta = -35\tfrac{1}{4}^\circ$ and at $\theta = +49^\circ$ corresponding to the AT- and BT-cuts, respectively.

The position of the two cuts relative to the major and minor rhomb faces, R and r, are shown in Fig. 8.2.

We have derived the expression for T_f using the room-temperature values of the elastic moduli and their temperature coefficients, and the linear coefficient of expansion. We therefore obtain the value of T_f for room temperature. If all the coefficients were constants and independent of temperature, then T_f would be independent of temperature.

However, none of the quantities is truly independent of temperature and therefore T_f must also be temperature-dependent. Bechman[5] has determined the second- and third-order terms in the power-series

[5] *Proc. I. R. E.,* **50**:1812 (August, 1962).

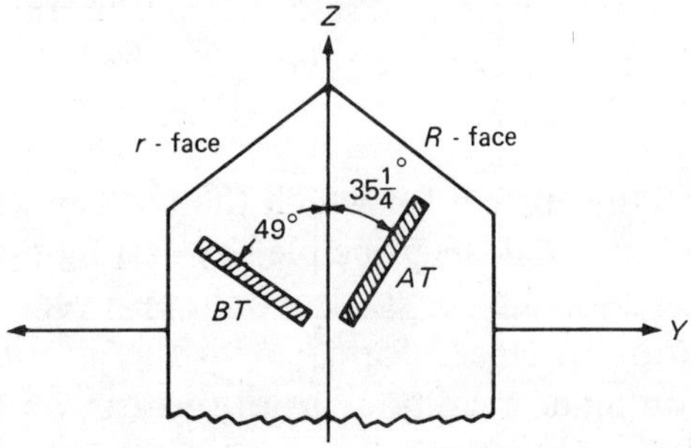

Fig. 8.2. The AT- and BT-cuts in quartz relative to the major and minor rhomb faces. The *X*-axis is normal to the plane of the diagram. The *r*- and *R*-faces make an angle of 38° 13′ with the *Z*-axis.

expression for the elastic moduli of quartz. Using these values it is possible, in principle, to calculate the value of T_f for any member of the rotated *Y*-cut family of piezoids at any temperature between -196 and +170°C.

Excellent experimental data exist showing the temperature-frequency curves for the AT- and BT-cut plates for the range of angles over which these cuts are used. Figure 8.3, which is adapted from the Final Report of the Union Thermoelectric Company on Contract No. DA49-025-sc-118, shows the temperature-frequency curves for AT-cut piezoids operating at 100 MHz on the fifth-harmonic mode. The blanks were 0.375 in in diameter and 0.0033 in thick, giving them a diameter/thickness ratio of more than 100. Their behavior, therefore, closely approximates that of an infinite plate.

Figure 8.4 shows the frequency-temperature curves of AT-cut plates operating at a frequency of 4.0 MHz. The blanks had diameters of 0.550 in and thicknesses of 0.0166 in, making the diameter/thickness ratio approximately 33. The blanks were designed for optimum performance by making the major surfaces conform to a spherical surface having a radius of curvature of 0.5 m.

From the curves of Figs. 8.3 and 8.4 it is apparent that no single angle can be prescribed for the AT-cut piezoid. The optimum value of the angle θ depends on (1) the nominal frequency, (2) the harmonic order, (3) the diameter/thickness ratio, (4) the temperature range, and to a lesser degree on (5) the size of the electrodes, (6) the drive level, (7) the countour, and (8) the material and thickness of the plated elec-

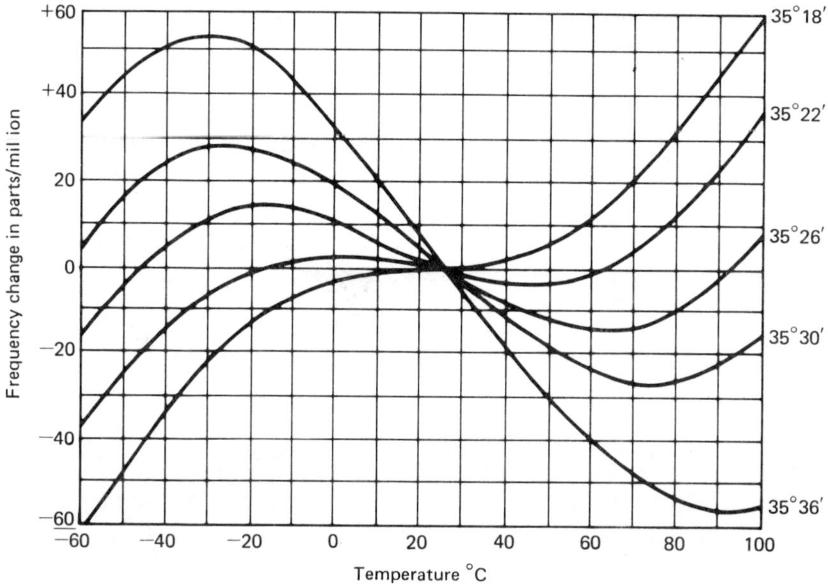

Fig. 8.3. Frequency-temperature curves for the AT-cut operating on the fifth harmonic mode at 100 MHz vs. the orientation angle.

trodes. Consequently, some experimentation is usually necessary to determine the correct value of θ and the allowable tolerances.

The temperature-frequency curves for the BT- cut are shown in Fig. 8.5. The frequency of a BT-cut plate cut at an angle of 49° has a maximum value near room temperature and the frequency decreases both at higher and lower temperatures. Consequently, the frequency is set above the nominal frequency at room temperature. Whereas an AT-cut plate can be made with a frequency tolerance of ±15 ppm over the temperature range −60 to +100°C, the BT-cut plate deviates by ±100 ppm over the same temperature range. For this reason the BT-cut has fallen into disfavor although it was formerly used very extensively.

However, the BT-cut does have certain advantages over the AT-cut. The frequency constant is 50 percent greater and therefore the blanks are 50 percent thicker for a given frequency. This is a distinct advantage at the highest frequencies. The turning point of the BT- but is easily adjusted to any desired temperature and the angular tolerances are less severe, so that the BT-cut has some advantages in applications in which the temperature of the crystal unit is controlled. Activity dips

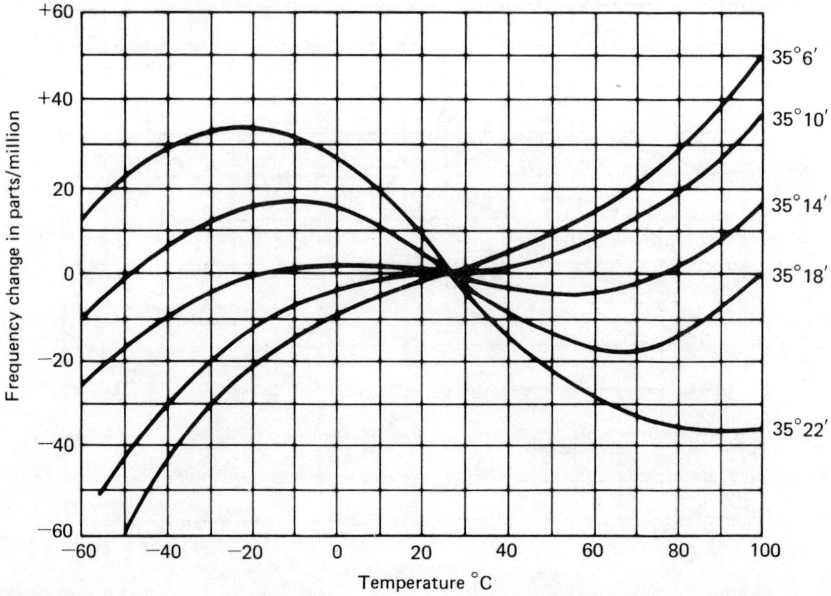

Fig. 8.4. Frequency-temperature curves for the AT-cut operating at 4 MHz vs. the angle of orientation.

are a bit less severe in BT- than in AT-cuts because the piezoelectric constant is smaller. The higher capacitance ratio, larger motional inductance, and smaller piezoelectric coupling sometimes offer advantages in filter applications.

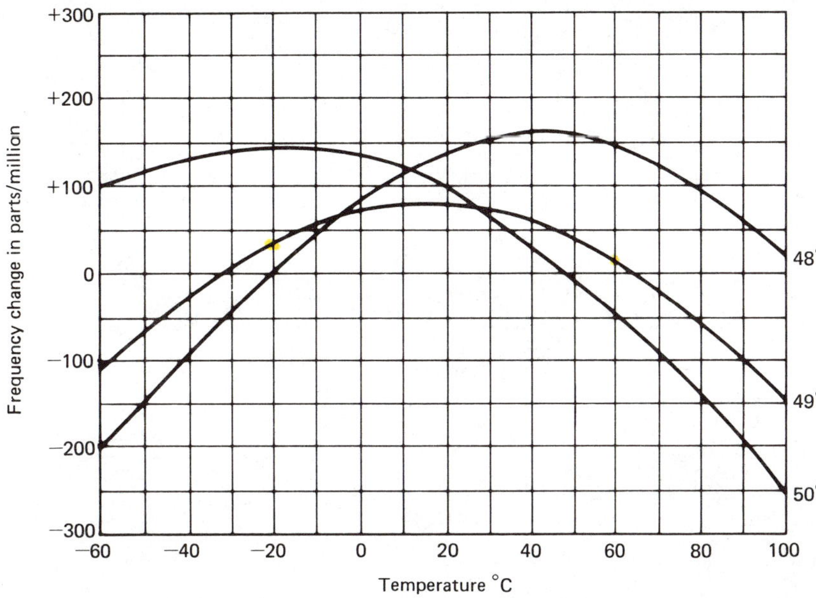

Fig. 8.5. Frequency-temperature curves for the BT-cut at different values of the angle θ.

9
COUPLED MODES AND ACTIVITY DIPS

ACTIVITY

The amplitude of oscillation of a crystal-controlled oscillator is usually determined, at least partially, by the characteristics of the crystal unit. If the oscillator is required to develop a certain minimum signal voltage, and if the equivalent resistance of the crystal unit is too high, the crystal unit may fail at one or more temperatures to oscillate at the specified amplitude. The grid current of the oscillator tube (or its solid-state counterpart) provides a simple measure of the amplitude of oscillation. The amplitude of oscillation, in a given circuit, under specified conditions, is often termed the *activity*. Crystal units were formerly specified in terms of activity, but since the invention of the CI-meter, the equivalent resistance has been used instead of activity since the resistance is characteristic of the crystal unit whereas the activity depends on both the crystal unit and the associated circuit.

ACTIVITY DIPS

Quite often the activity of a crystal unit decreases abruptly or even drops to zero at a certain temperature. Such a decrease is called an *activity dip.* Since it is much easier to monitor a quantity such as activity than to monitor the equivalent resistance, the activity is still often used in testing crystal units. Correlation between the two quantities is easily established. All the motional parameters of the equivalent circuit undergo changes near the temperature at which the activity dip occurs. The equivalent resistance increases (decreasing the amplitude of oscillation) and the Q decreases while L, C, and their product, which determines the resonant frequency, deviate from their normal values.

Figure 9.1 shows the variation of activity and frequency with temperature of a 5-MHz AT-cut plate. An activity dip occurs at a temperature of about 50°C in this particular unit.

The activity and frequency of a crystal unit may vary abruptly at some temperature for several different reasons. A unit encapsulated in an atmosphere containing water vapor usually displays a decrease in activity and frequency as the temperature of the crystal unit is raised through the dew point. Condensation of water on the cold blank produces both loading and damping, resulting in the observed changes. This phenomenon has been used as a dew-point indicator. Contaminants such as solder flux and poorly adhering plating sometimes also produce similar perturbations of the activity and frequency at certain temperatures. Piezoids made from cultured quartz sometimes show reduced activity at low temperatures, but this is not a true activity dip and in any case is not very important, since the quality of cultured quartz has been greatly improved with experience in production.

True activity dips are caused by coupling between the desired mode of vibration and some other mode of vibration which happens to have the same frequency at that temperature. We have already discussed some of the modes of vibration which can be excited in an

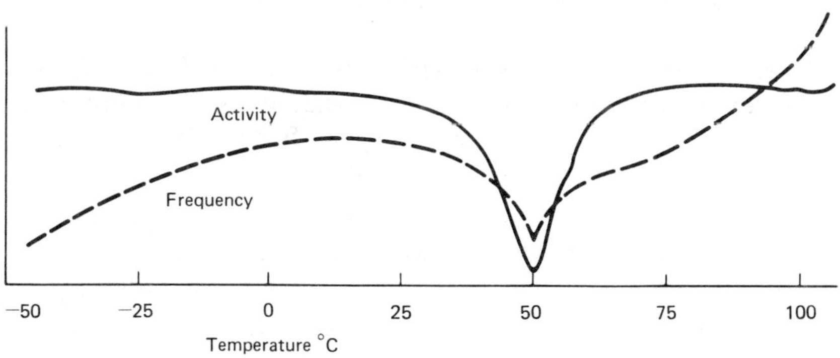

Fig. 9.1. Frequency and activity of a 5-MHz AT-cut plate vs. the operating temperature (experimental).

elastic solid. In addition to the inharmonic modes of the thickness-shear vibration, we also have the face-shear mode and all its overtones; longitudinal vibrations which may be excited under certain conditions; and, perhaps most serious of all, flexure modes which sometimes occur. In addition to the various modes of vibration of the quartz plate, vibrations of various kinds may be excited in the mounting structures, and in low-frequency piezoids, even air resonances may occur. When one of these modes is excited energy is thereby dissipated, thus increasing the equivalent resistance R and decreasing Q. The mass and/or stiffness added to the vibrating system causes changes in the frequency of resonance. The interaction between the two modes of vibration is called *coupling*.

COUPLING IN VIBRATING SYSTEMS

Coupling between mechanical or electric oscillating systems is a common physical phenomenon. Figure 9.2 shows a simple experiment for demonstrating the principles involved. Two simple pendulums, P_1, and P_2, consisting of metal spheres and fine threads are supported from a tightly stretched horizontal cord. When one of the pendulums is set into vibration, part of its energy is transferred to the other, thereby setting it into motion. If the lengths of the two pendulums are adjusted so that their frequencies are equal, all the energy of the first pendulum will be transferred to the second pendulum after a few cycles. If the frequency of the second pendulum differs from that of the first, little or no motion is excited in the second pendulum.

If the motion of one of the pendulums is heavily damped, by submerging it in water or some viscuous liquid, and if the resonant frequencies are the same, the vibration of the first pendulum is quickly

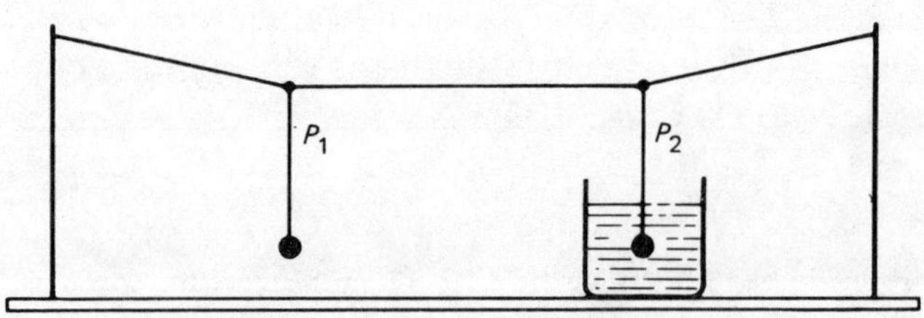

Fig. 9.2. Coupled pendulums with P_2 damped.

damped out. If the frequencies are quite different, little energy is transferred to the damped pendulum and therefore the motion of the first is only slightly affected.

The coupling in quartz crystal units may be either mechanical or electrical or a combination of the two. We first look at examples of coupling in which the electric field excites other modes of vibration in addition to the desired mode. Then we examine some examples of mechanical coupling and finally some of the more complex cases in which both are involved.

COUPLING BY COMMON ELECTRIC EXCITATION

To become familiar with the principles involved we first consider the simple case of an X-cut plate with the electrodes producing an electric field in the X-direction. The piezoelectric strain matrix for quartz shows [see Eq. (26)] that an electric field in the X-direction produces three types of strain. They are a longitudinal strain in the X- or thickness direction, a longitudinal strain in the Y-direction, and a shear strain in the YZ-plane, i.e., about the X-axis. If the electric field is due to an alternating current of the correct frequency, the piezoid may be set into vibration in one or more of the following ways. A longitudinal wave may be excited in the thickness direction excited through the piezoelectric coefficient d_{11}. A longitudinal wave may be excited in the Y-direction through d_{12}. A face-shear vibration may be set up in the plate (in the YZ-plane) through piezoelectric coefficient d_{14}. Harmonic and inharmonic overtone modes of each of the modes may also be excited, making the number of modes which can be excited very large and distributing them over a large range of frequencies. The temperature-frequency coefficients of the modes are unequal, since they are controlled by different elastic coefficients, so the possibility exists that the frequencies of two or more modes may become equal at certain temperatures. An activity dip occurs when the frequency of one of the other modes coincides with that of the desired mode. The crystal unit behaves electrically like the circuit of Fig. 9.3, where we may consider the parameters L' and C' to be temperature-dependent.

Consider a Y-cut plate with electrodes producing a field in the Y-direction. The piezoelectric matrix for quartz, Eq. (26), shows that two types of strain are produced by the field in the Y-direction. The

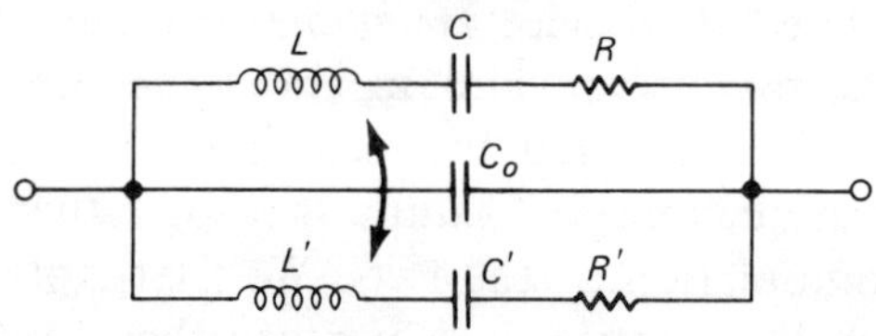

Fig. 9.3. Equivalent circuit of two modes excited by the same electric field. Coupling exists between *L* and *L'*.

coefficient $d_{25} = d_{14}$ relates a shear strain in the *XZ*-plane with the field in the *Y* direction. If the field varies at the correct frequency, the plate may be set into vibration in one of its many face-shear modes.

The coefficient $d_{26} = -2d_{11}$ relates a shear strain in the *XY*-plane to the field in the *Y*-direction. Again, if the field is produced by an alternating current of the proper frequency, a shear wave is excited which travels in the thickness direction of the plate. This is, of course, the mode of motion employed in the AT- and BT-cut plates. This mode also has many harmonic and inharmonic overtones which may be excited, as we have previously shown. The temperature coefficients of all these modes is substantially zero.

If the normal to the plate is not exactly parallel to the *Y*-axis, the electric field has a component in either the *X*- or *Z*-direction or both. Fortunately, the *Z*-component does not excite any piezoelectric strain, but a component in the *X*-direction may excite any of the modes which can be excited in an *X*-cut plate. Since an AT- or BT-cut plate is simply a *Y*-cut plate rotated about the *X*-axis, any field component in the *X*-direction is capable of exciting these same modes in these plates. For this reason it is important in fabricating AT- and BT-cut plates to be sure that the normal to the plate lies in the *YZ*-plane.

COUPLING BY MECHANICAL OR ELASTIC EXCITATION

In addition to the many modes of vibration which can be excited by the same electric field, other modes can be excited through a common strain. As an example of such a mechanically coupled system we may consider a bar of quartz with its rectangular cross section in the *XZ*-

plane and its length in the Y-direction. Many frequency standards were formerly made in this way. The field is applied in the X-direction and the bar is set into longitudinal vibration in the Y-direction through the coefficient d_{12}. The resulting extensional strain, y_y, produces extensional stresses in the X- and Z-directions through the stiffness coefficients c_{12} and c_{32} of Eqs. (13 and 13b), and a shear stress in the YZ-plane through coefficient c_{42} (which is equal to $-c_{14}$).

We consider first the extensional stress in the Z-direction produced through the elastic coefficient c_{32}. Although it is impossible to excite a longitudinal vibration in the Z-direction in quartz by an electric field, such a vibration can be excited by coupling to the extensional vibration in the Y-direction through the cross-coupling coefficient c_{32}. If the two frequencies are compatible, a standing wave in the Z-direction may be excited by mechanical coupling with the basic vibration consisting of a standing wave in the Y-direction.

In the same way, the extensional stress in the X-direction, X_x, produced through the coefficient d_{11} may excite longitudinal vibration in the Y-direction. This driving force is in addition to that produced by the electric field which is, of course, able to excite the same motion.

Finally, the Y_z stress produced through the coefficient c_{42} can produce a face-shear resonance if the frequencies are compatible and if the boundary conditions permit. This mode, like the previous one, can also be excited electrically so that the mechanical coupling is added to the electric coupling.

A particularly interesting and important example of coupling is that of the coupling between the face-shear modes and the thickness-shear modes in the Y-family of quartz crystal units. It is somewhat easier to think first of a Y-cut plate and later extend the ideas to the AT- and BT-cut plates. The thickness-shear mode of the Y-cut involves an x_y strain, but from Eq. (13b) it follows that an x_y strain produces a $-Z_x$ stress through the cross-coupling coefficient c_{56}. Stress Z_x produces strain z_x, which is the strain involved in the face-shear mode of vibration. The result is that the face-shear vibration is excited by the thickness-shear vibration through the cross-coupling coefficient c_{56}; and if the conditions are correct, resonance may occur. The mechanical coupling between the thickness- and face-shear modes thus augments the electric coupling which exists since both modes may also be excited by the same electric field. The result is that the two modes are very strongly coupled and the frequency constant is perturbed in the manner shown in Fig. 9.4.

We consider next an AT-cut plate having lateral dimensions 1.0 × 1.0 cm and thickness 0.60 mm. The thickness-shear frequency is approximately 2.8 MHz and the frequency of the 11 face-shear mode is about 300 kHz. The frequency of the 99 overtone of the face-shear vibration is very nearly equal to that of the thickness-shear mode, and since the frequency of the face-shear mode changes with temperature, the two may become equal at some point within the specified temperature range.

The coupling between the thickness- and face-shear modes in the AT-cut plate is provided by the cross-coupling coefficient c'_{56}, which may be calculated by the same method used earlier to calculate c'_{66}. The result is found to be

$$c'_{56} = c_{14}\,(c^2 - s^2) - (c_{66} - c_{44})\,sc$$

which becomes zero for two angles,

$$\theta = 31°46' \text{ and } \theta = -58°15'$$

Plates cut at these angles are designated the AC- and BC-cuts, respectively. The C signifies that the coupling is zero. The AC-cut has been used as a thermometer. It has a T_f of +20 ppm/°C and its freedom from coupling to face-shear modes makes it relatively free from frequency perturbations.

The value of c'_{56} is small at the orientation of the AT-cut, which is only 3.5° from the AC-cut. Consequently, the coupling between the thickness-shear and face-shear modes is small. The coupling be-

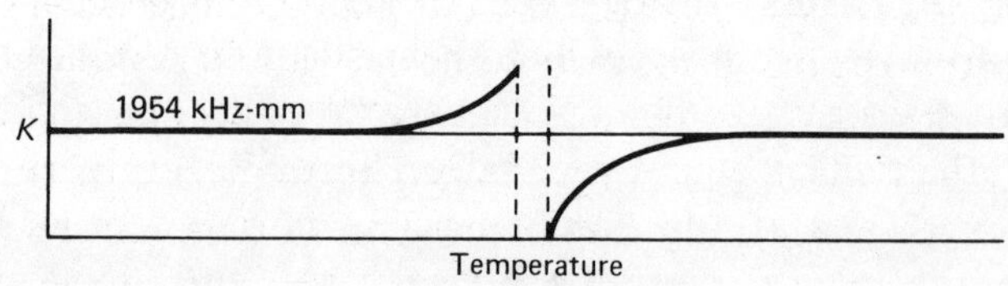

Fig. 9.4. Perturbation of the frequency constant by coupled modes in the *Y*-cut plate.

tween the two modes is greater in the BT-cut, which is over 9° from the BC-cut. It is unfortunate that the angles for zero coupling and zero T_f are not the same.

VIBRATIONS IN MOUNTING STRUCTURES

The piezoid must always be supported in some manner. Unless the mounting structure can be attached at points which are nodes of motion, the mounting structure becomes a part of the vibrating system. This phenomenon is particularly troublesome in CT- and DT-cut piezoids vibrating in the face-shear mode. The supporting wires are usually attached at the nodal points at the center of the plates. If the point of attachment is not exactly at the nodal point, the supporting wire becomes a part of the vibrating system and at certain temperatures may be set into resonance, thereby absorbing energy and causing an activity dip.

In AT-cut plates the vibrating area is usually confined to the center of the plate by making the electrodes considerably smaller than the plate. The zone around the edge may then be used to support the plate. If the lateral dimensions of the plate are large compared with its thickness, little difficulty is encountered with damping due to energy transferred to the mounting structure. If the ratio of lateral dimensions to thickness is less than about 20, it is difficult to find nodal points at which to support the plate. Points which are nodal points at one temperature become antinodes at another temperature; the mounting structure is set into vibration and an activity dip results.

COUPLING THROUGH COMMON BOUNDARY CONDITIONS

A transverse wave traveling in the direction of a lateral dimension of a plate is called a flexure mode. These modes are often excited in thickness-shear plates such as the AT-cut plate because of the similarity between the strains in the two modes. Figure 9.5 shows the type of strains involved in the thickness-shear vibration of an AT-cut plate and a flexure mode in the same plate. The similarity of the strains at the ends is immediately apparent. The flexure modes are very easily excited in rectangular AT-cut plates, especially if the length of the plate in the X-direction is less than 20 to 30 times the thickness.

Fig. 9.5. Edge view of the strains in an AT-cut plate (*a*) vibrating in thickness shear and (*b*) vibrating in its first even flexure mode.

Some of the modes of motion of a bar or plate vibrating in flexure are shown in Fig. 9.6. Only the even-order modes are strongly coupled. Neither the strains at the boundaries nor the dynamical balance is conducive to coupling between the thickness-shear and odd-order flexure modes.

The thickness-shear mode tends to excite a transverse wave which travels in the *X*-direction in the blank. The frequency of the thickness-shear vibration is substantially independent of the temperature but that of the flexure mode is not, so that at some temperature the frequency of the thickness-shear mode is equal to that of one of the flexure modes and the plate is set into resonance with severe effects on both the frequency and the activity of the plate. A dramatic demonstration of the coupling to the flexure mode may be made by using an AT-cut plate having dimensions of 15 to 20 mm and a frequency of 2 to 3 MHz. The rectangular plate is mounted between parallel electrodes with a small air gap and excited at its resonant frequency in a circuit which drives it strongly. By adjusting the length in the *X*-direction, or the temperature, or both, conditions for coupling to one of the even-order flexure modes can be established.

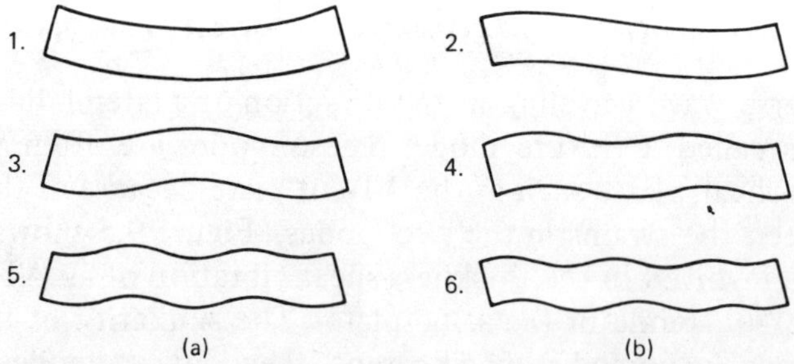

Fig. 9.6. Edge view of the displacement in plates vibrating in (*a*) odd flexure modes and (*b*) in even flexure modes.

If the space between the electrode and the quartz blank is observed in a dark room while the blank is strongly excited, a glow discharge may be observed in the form of parallel lines perpendicular to the X-direction. The strain is apparently sufficient at the antinodes of the flexure to produce voltages great enough to produce ionization of the air and the glow which results. If operated in this manner for a little time, the electrodes will be pitted with the pits outlining the position of the nodes and antinodes of the flexure motion.

Flexure modes are most troublesome in rectangular plates and at low frequencies. This is one of the reasons for preferring circular plates over rectangular plates in the fabrication of AT-cut piezoids.

It is possible to utilize the coupling between the flexure modes and the thickness-shear mode to obtain excellent temperature coefficients over a limited range of temperature. The thickness constant K is changed by the strong coupling between the two modes and if the lateral dimensions of the plate are less than about 10 times the thickness, the change may easily be ± 10 percent. Such plates are very difficult to mount, so their application is limited.

EFFECTS OF TEMPERATURE

Because of the various types of coupling which may exist, a crystal unit may be an excellent resonator at one temperature and quite inactive at another temperature. Sometimes the frequency "skips" from one value to another as the temperature is changed. The reason for this behavior is to be found in the different frequency-temperature coefficients of the various modes. The piezoid is usually designed to make the value of T_f as low as possible, but it is not possible to make T_f for all of the other modes simultaneously also zero. Thus at some temperature the frequency of another mode may become equal to that of the resonator and the activity and frequency are thereby altered.

The inharmonic modes of the thickness shear do not contribute to the problem of activity dips. They have the same frequency-temperature coefficient as the main mode and therefore cannot interfere. It does happen sometimes that as the activity of the main mode is degraded by an interfering mode, the oscillator frequency "jumps" to that of one of the inharmonic modes to return to the main mode at another temperature.

DIMENSIONING

The frequency of the main mode in quartz crystal units employing AT- and BT-cut plates is determined by the thickness of the plate. The frequencies of most of the interfering modes are determined by the lateral dimensions. Therefore, the lateral dimensions must be tightly controlled to avoid coupling between desired and undesired modes of motion. The process of adjusting the lateral dimensions to avoid coupled modes is called *dimensioning.* Crystal units made with square or rectangular blanks must be dimensioned very carefully to avoid activity dips and to obtain a reasonable production yield.

Most of the coupled modes are much less severe in circular than in rectangular blanks. Consequently, most crystal units in the higher-frequency range are made with circular blanks. The face-shear mode, however, can still be excited in a circular blank; and in applications in which the entire blank is excited by the electric field, the diameter of the blank becomes important. Such modes cannot exist if only the center portion of the blank is excited, because the active region is surrounded by a zone of inactive quartz. If the width of this zone is 20 or more times the thickness of the blank, it also provides an inactive region for supporting the blank.

LOW DIMENSIONAL RATIOS

An infinitely long bar or an infinitely large plate has only a single mode of vibration and hence activity dips due to coupled modes would never appear. In actual plates the severity of the activity-dip problem depends upon the lateral/thickness dimensional ratio and becomes rapidly more severe as the ratio drops below about 20. Thus the activity-dip problem may be quite insignificant in a 10-MHz AT-cut plate having a diameter of 0.75 cm in which $d/e = 45$ and very troublesome in a 2.5-MHz unit in which the AT-cut blank has a diameter of 1.00 cm and a dimensional ratio of 15.

The use of AT-cut blanks in which d/e is less than about 20 should be avoided where possible to avoid the difficulty of producing units of this kind which are free of activity dips over an extended frequency range and, of course, the unavoidable economic cost.

The new SC-cut (Chap. 13) is relatively free of activity dips and for this reason may find application in certain cases where activity dips are a serious problem with AT-cut plates.

Another consequence of a low *d*/*e* ratio in an AT-cut plate is that the product of the frequency and the thickness is not a constant (page 135). This is due to the coupling between the thickness-shear mode and low-order overtones of other modes which strongly perturbs the vibration frequency. The frequency-temperature curve is distorted for the same reason. The familiar "cubic" curve of the high-frequency AT-cut resonator may be entirely unrecognizable in a low-frequency unit having a *d*/*e* ratio of 10 or less.

If units must be designed with *d*/*e* ratios less than about 20, it must be expected that much effort will be required to find a suitable design and much care will be required to hold the required tolerances in production.

10
INHARMONIC OVERTONE MODES AND FILTER CRYSTALS

INTRODUCTION

The use of thickness-shear resonators as components in HF and VHF wave filters was delayed for many years because of the existence of the inharmonic overtone modes which are present in all such piezoids. As we have seen in Chap. 5, the frequencies of these modes lie in a relatively narrow range immediately above that of the main mode where they cause severe problems in the design of wave filters. These modes came to be called *spurious modes* and later *spurs.*

It is the purpose of this chapter to review the nature of these modes and to discuss the methods which are available for dealing with them in the fabrication of crystal units for use in HF and VHF filters.

Before continuing with this chapter the reader may find it useful to review the material presented in Chap. 5.

NATURE OF THE INHARMONIC OVERTONE MODES

The inharmonic overtone modes, or spurs, are not an expression of the perversity of nature or of demon possession. On the contrary they are natural phenomena just as are the overtones of an organ pipe or piano string. They are more closely analogous to the overtones of a bell which, in general, are all higher in pitch than the fundamental and have frequencies which are not harmonically related to the fundamental or to each other.

The overtones of any vibrating system are determined primarily by the nature of the medium and the boundary conditions and very little by the method of excitation. For example, a violin string tuned to the note G produces this note whether the string is plucked or bowed or even if the violin is dropped on the floor. It is most important to note this point. It is especially applicable to quartz resonators in which the damping losses are so small that any mode which is physically possible and permitted by the boundary conditions may be (and probably will be) excited.

It should be recalled that inharmonic modes do not occur in one-dimensional vibrating systems in which the only possible modes are the harmonic overtone modes. In two- and three-dimensional systems inharmonic as well as harmonic modes are possible. An infinitely large AT-cut plate with electrodes covering the entire surface would behave as a one-dimensional resonator and would have only harmonic modes with $n = 1, 3, 5, \ldots$. In a finite plate different sets of inharmonic overtones are produced depending on whether the edges are clamped or free. If the edges are clamped, nodes of motion must occur at the edges while a free edge acts as an antinode of motion.

THE GEOMETRY OF THE PIEZOID

Since we are here concerned with the overtone modes of the resonators used in HF and VHF wave filters, we shall consider a typical unit consisting of a circular AT-cut plate. The electrodes are deposited on the quartz in the form of a circular area concentric with the boundary of the plate. The plate is assumed to have plane, parallel surfaces although in most cases the surfaces are slightly convex due to the nature of the lapping process used in fabricating the blanks.

Except for a small amount of fringing at the edge of the electrodes, the electric field is confined to the region between the electrodes and the piezoelectric strain is therefore confined to this portion of the plate. The zone of quartz surrounding the electrodes is not excited piezoelectrically, so we may think of it as a device for supporting the vibrating area much as the frame of a drum supports the stretched membrane which forms the drumhead. We will see later that the surrounding zone plays an important role in determining the characteristics of the resonator.

THE FUNDAMENTAL, OR *n*01, MODE

The primary excitation of the AT-cut plate consists of the $X_{y'}$ shear strain. The resulting vibration at the resonant frequency consists of a transverse wave traveling in the thickness direction of the plate. The particle motion is in the *X*-direction. The plane parallel to and midway between the surfaces is a nodal plane and consequently the displacement on opposite sides of it are in opposite directions.

At the surface of a vibrating plate the mechanical displacement and the electric polarization have similar spacial distributions. This may be a bit confusing, since the polarization depends on the strain and not the displacement. It must be remembered, however, that the polarization depends on the strain *in* the plate rather than the strain at the surface (which is zero) and that the internal strain is greatest at the center of the plate. The displacement, strain, and polarization all become zero near the edge of the electrode.

Numerous experiments show that the amplitude of the motion decreases substantially to zero at a distance less than 20 times the plate thickness from the edge of the electrode. It is likely that a part of the excitation in the zone immediately surrounding the electrodes is due to fringing of the field. Figure 10.1 gives some idea of

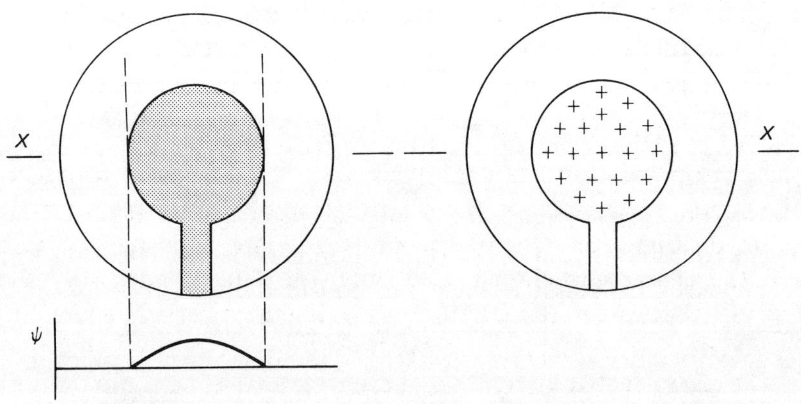

Fig. 10.1. The displacement and charge distribution which would result if an unelectroded plate were excited in the *n*01 mode.

the distribution of the displacement across the diameter of the plate in the direction of the X-axis.

If the plate were infinitely large and the driving field were uniform over the entire surface, the distribution of the strain and the surface charge density would be constant over the entire surface. The motion would be that of a plane wave traveling in the thickness direction of the plate. In an actual resonator the strain must decrease to zero at some point not far from the edge of the electrode area. Furthermore, the conducting electrodes make the surface charge density approximately constant over the surface. This results in an apparent increase in the elastic constant and tends to raise the frequency of the resonator. Since this effect is small, we will ignore it along with the effects of anisotropy.

The distribution of motion over the vibrating area may be considered to result from a wave traveling radially on the surface and being reflected at the boundary between the vibrating and nonvibrating regions. The result is a standing wave with a nodal line around the circumference and an antinode in the center. The standing wave is the result of interference between waves incident on, and reflected from, the edge in the same way that standing waves are set up on a string by interference between the waves traveling in opposite directions on the string. A closer analogy is that of the vibration of a circular drumhead in which the center of the drumhead moves in a direction perpendicular to its plane while the edge remains stationary. (It must be emphasized that this is only an analogy. The motions are quite different in the two cases.)

THE FIRST-INHARMONIC OVERTONE, OR n11, MODE

When the plate is excited in the $n11$ mode, waves travel around the plate in opposite directions, setting up a standing wave with a nodal line along a diameter of the vibrating area. The displacement and strain on opposite sides of the nodal line are opposite in direction and the charges developed have opposite signs. Figure 10.2 shows the distribution of the strain along a diameter perpendicular to the nodal line. The nodal line around the edge is, of course, still present.

The piezoelectric current to the resonator is the algebraic sum of the currents to the two regions and the fact that these two currents are 180° out of phase suggests that if the strains, their distribution,

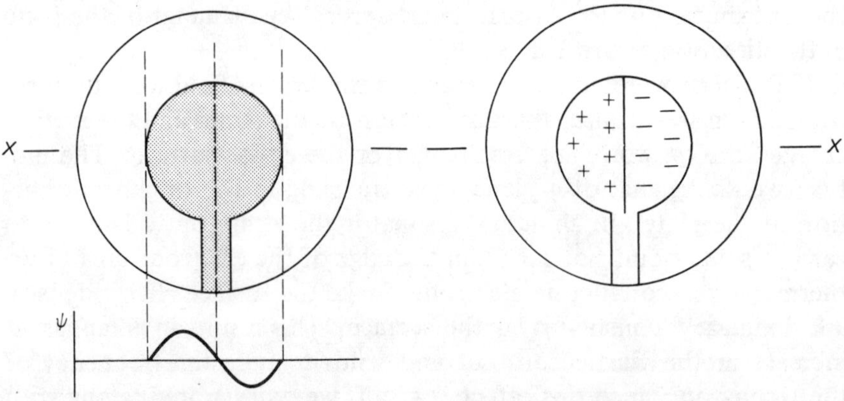

Fig. 10.2. The displacement and charge distribution on the surface of an unelectroded AT-cut plate vibrating in the 111 mode.

and the areas of the two regions were identical, then the piezoelectric current should be zero and the mode should not be excited. If excitation of the mode were desired, it could be achieved by dividing the plating at the nodal line and crisscrossing the connections so that the fields would be opposite in the two portions.

In practice the quartz plate should be lapped to have as nearly perfect symmetry as possible and the electrodes should be carefully aligned and located concentric with the blank. This is important because the $n11$ mode is the first overtone above the fundamental and is therefore the most troublesome one. Although it is possible, in theory, to suppress the $nm1$ modes by careful control of symmetry, and it is true that the modes are often missing or very weakly excited, it is difficult to eliminate them entirely in this way. The slightest asymmetry is sufficient to cause the plate to be excited in one of these modes.

THE $n02$ MODE

The second-overtone mode of a circular plate is the $n02$ mode. When this mode is excited a radial standing wave is set up. The distance from the center of the plate to the edge is one wavelength and consequently a second circular nodal line is formed concentric with the nodal line at the circumference. This divides the surface of the plate into two zones. Naturally the displacement, strain, and polarization are opposite in the two zones. The displacement along any diameter of the plate is shown in Fig. 10.3.

It would be difficult or impossible to suppress the *n*02 mode or any other mode having circular nodal lines by controlling the symmetry of the quartz blank and the plating, since the strains and charges also have circular symmetry.

THE EFFECTS OF CONTOURING

The frequency separation between the main and overtone modes is inversely proportional to the square of the radius, or the area, of the vibrating region, as shown in Eq. (42). Therefore, reducing the size of the vibrating area should cause the troublesome inharmonic modes to move to higher frequencies, which is indeed the case.

One method of reducing the size of the vibrating area is to make the blank thinner at the edges than in the center. This is usually done by lapping one or both sides to have a spherical shape. The process of generating a spherical surface on a quartz plate is called *contouring.* Spherical surfaces are used, not because they are necessarily the best configuration, but because spherical surfaces are relatively easy to generate using the techniques for grinding spherical surfaces on lenses.

As the radius of curvature of the surface of the quartz plate is reduced, the frequencies of the inharmonic overtone modes move to higher frequencies and the overtone spectrum becomes more dispersed.

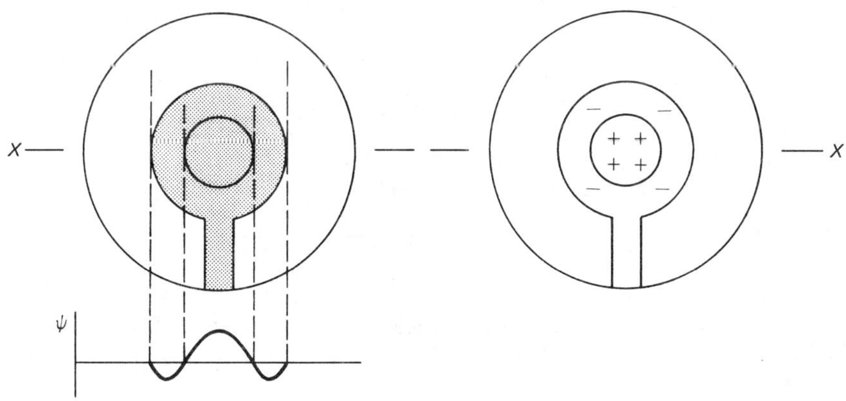

Fig. 10.3. The displacement and charge distribution on the surface of a typical AT-cut plate excited in the *n*02 mode.

It is easy to increase the frequency interval between the main mode and the first inharmonic overtone mode by 2 percent or more of the nominal frequency.

Figure 10.4 shows the effect of contouring on the overtone frequency spectrum of a circular AT-cut plate having a diameter of 0.550 in (1.4 cm). Initially the surfaces of the plate were plane and parallel. One side of the blank was contoured using an optical lap to generate a spherical surface. The entire surface of the blank can be made spherical without changing the frequency of the fundamental mode if care is taken not to reduce the thickness of the plate at the center.

Laps with successively smaller radii of curvature were used. The ordinate of Fig. 10.4 is given in units of $1/2R$, where R is the radius of curvature of the lap measured in meters. (This corresponds approximately to the diopter rating used by opticians.) An ordinate of 1 corresponds to a radius of curvature of 0.5 m, an ordinate of 2 corresponds to a radius of 0.25 m, etc.

The effect of decreasing the radius of curvature of the spherical surface is readily apparent. Initially, when the surface is "flat" the

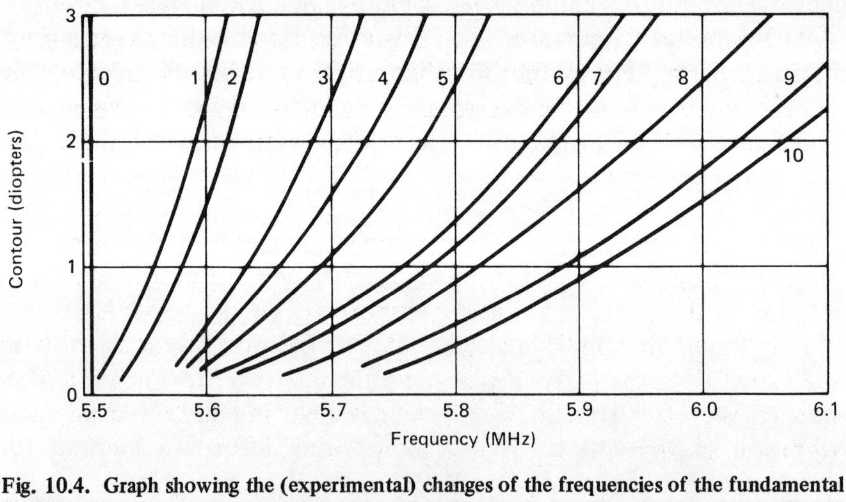

Fig. 10.4. Graph showing the (experimental) changes of the frequencies of the fundamental and first 10 overtones of a 5.5-MHz AT-cut blank as the radius of curvature of one side was changed from infinity (a flat blank) to 0.16 m. The ordinate is in units of $1/2R$, where R is in meters. The numbers are roughly equal to the so-called diopter reading which is marked on optical laps.

frequencies of the overtone modes are very closely grouped just above the frequency of the main mode. The frequencies are so close together that their resonances tend to overlap and it is impossible to determine the resonant frequencies uniquely. As contouring proceeds, the frequency of each overtone mode moves higher in a quite regular way. Dust-pattern studies show that each mode retains its characteristic identity and that the area of the vibrating portion of the blank decreases as contouring proceeds. The vibrating areas may be either smaller or larger than the electrode area and the nodal pattern is independent of both the size and location of the electrode, although the relative amplitudes or equivalent resistances of the modes is thereby affected.

Other effects accompany the changes in the overtone spectrum as the blank is contoured. Slight contouring reduces the value of the equivalent resistance R by reducing damping losses through the mounting supports and by moving the spurious modes away from the resonant frequency. Further contouring causes the value of R to increase because of the decrease in the vibrating area.

The values of L and C are also affected. As the contouring proceeds L increases and C decreases. Hence, contouring provides a method for controlling the values of these two parameters.

The capacitance ratio $r = C_0/C$ is also influenced by contouring. It is shown in App. IV that the theoretical value of r for a large, thin AT-cut plate should be about 154. In arriving at this value it is assumed that the distribution of strain is uniform over the entire electroded area and that no fringing occurs at the electrode edge. Since neither of these conditions is satisfied and because the holder adds some capacitance to C_0, the measured value of r is always greater than 154. By careful attention to the size of the electroded area in relation to the contour, values of r as low as 190 can be achieved.

It is shown in App. IV that the value of r is closely associated with the value of κ, the piezoelectric coupling coefficient, which is a measure of the effectiveness of the electric field in exciting mechanical vibration in the quartz. Hence, contouring affords a method for changing the value of κ.

Although contouring does provide a degree of control over the values of L and C and the frequencies of the overtone modes, its usefulness is limited by the fact that the two effects are not independent.

The contour required to disperse the overtone modes may not yield the values of L and C which are simultaneously required.

Contouring is most useful in the range of frequencies between about 2 and 5 MHz. At higher frequencies, 10 to 100 MHz, which is the range in which many applications occur, contouring is impractical or at least a very difficult production process.

Fortunately a method of eliminating or at least suppressing the undesired overtone modes without unduly influencing the motional inductance and capacitance is available and it is especially applicable at the higher frequencies. It is called *energy trapping.*

ENERGY TRAPPING

Resonance in a vibrating system is always associated with the phenomenon of standing waves. In order for standing waves to occur, certain simple relationships must exist between the dimensions of the resonator and the wavelength of the disturbance being propagated on it. For example, the length of a stretched string, vibrating in a resonant mode, must be an integral number multiplied by half the wavelength of the disturbance on the string. The wave which travels one way on the string is reflected at the end and the interference between the incident and reflected waves produces the standing waves and resonance. It is impossible for resonance to occur in a string of infinite length because the reflected wave would not return to the source in a finite time. In an actual vibrating system some energy is dissipated as the wave is propagated and consequently the wave may be dissipated before it returns. Such a system could not sustain standing waves. If the incident wave could be completely absorbed by some mechanism, then no energy would be reflected and again no standing wave would result.

Some waves travel in a tangential direction in circular and spherical vibrating systems. The condition for standing waves and resonance to occur in these cases is that the length of the path followed by the wave must be an integral number of wavelengths. Those modes of motion (see Fig. 5.8) having nodes along diameters are associated with waves traveling in the tangential direction. The number of nodal diameters is equal to the number of wavelengths divided into the length of the path. These waves are not reflected at any boundary but result from the interference of two waves traveling in opposite direction in the tangential directions.

On the other hand, resonance associated with waves traveling in the radial direction depend on reflection at the circular boundary. In this case the nodal lines are concentric with the center of the plate.

Both types of standing waves occur in circular plates. It is important to note that the modes which produce circular nodes are due to reflection of waves at the edge of the plate, but those modes which result in nodes along the diameters are not due to reflection but rather to the condition that the length of a closed path must be an integral number of wavelengths.

The theory of energy trapping is based on the principle that traveling waves can be propagated and thereby produce standing waves and resonances only for frequencies above a certain minimum value called the *cutoff* frequency. The fundamental mode frequency is the cutoff frequency of a one-dimensional system, but this is of only trivial interest.

As we have seen, in a multidimensional system some modes exist which have frequencies only a little above the fundamental mode. It can be shown that these modes may be trapped and therefore prevented from being reflected, thereby setting up the resonances which are called the inharmonic modes or spurs.

The mathematics of the problem for a system with circular symmetry and including the effects of anisotropy is very complex, but we can see the general features and even make some useful calculations by considering the propagation of waves in a much simpler system. For this purpose we consider the propagation of waves in an isotropic plate of thickness e and width w, oriented with respect to orthogonal axes, as shown in Fig. 10.5.

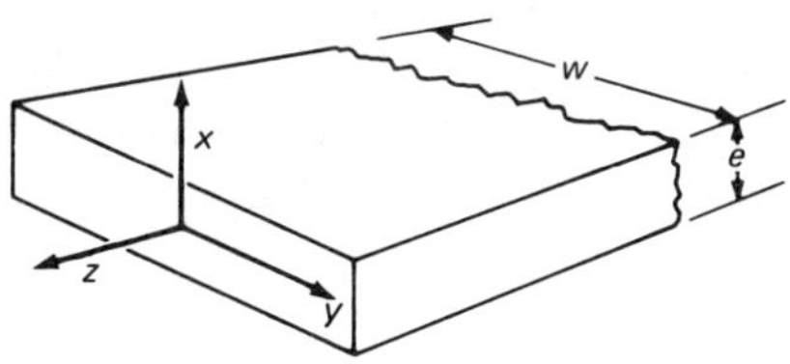

Fig. 10.5. Thin plate with waves propagated in the length or z direction. All surfaces are assumed to be free (not clamped).

The equation for the displacement, expressed in rectangular coordinates, is

$$\frac{\partial^2 \psi}{\partial t^2} = v^2 \left(\frac{\partial^2 \psi}{\partial x^2} + \frac{\partial^2 \psi}{\partial y^2} + \frac{\partial^2 \psi}{\partial z^2} \right) \tag{93}$$

We are interested in solutions of Eq. (93) which describe traveling waves in the z direction. Such waves, reflected at a boundary, produce standing waves which are associated with the inharmonic overtones.

Such a solution is

$$\psi = [X(x)]\,[Y(y)] \cdot [\exp j\,(\omega t \pm kz)] \tag{94}$$

where $k = 2\pi/\lambda_z$ and λ_z is the wavelength of the wave traveling in the z direction produced by the disturbance of frequency ω.

By computing the various second derivatives of Eq. (94) and substituting in Eq. (93) we have

$$k^2 - \frac{\omega^2}{v^2} = \frac{1}{X}\,\frac{d^2 X}{dx^2} + \frac{1}{Y}\,\frac{d^2 y}{dy^2}$$

Using the usual method of separation of variables

$$\frac{1}{X}\,\frac{d^2 X}{dx^2} = -\zeta^2 = k^2 - \frac{\omega^2}{v^2} - \frac{1}{Y}\,\frac{d^2 y}{dy^2} \tag{95}$$

where ζ^2 is a constant to be determined.

The solution of the equation involving x is

$$X = A\cos(\zeta x) + B\sin(\zeta x)$$

The boundary conditions require that $\psi = 0$ at $x = 0$ and to be a maximum at $x = \pm\frac{1}{2}\,e$. These conditions are satisfied if $A = 0$ and $\zeta = (n\,\pi/e)$, where $n = 1, 3, 5, \ldots$. The portion of Eq. (95) involving y can be written

$$\frac{d^2 y}{dy^2} = \gamma^2 Y \tag{96}$$

where

$$\gamma^2 = \zeta^2 + k^2 - \frac{\omega^2}{\nu^2} \tag{97}$$

The solution of Eq. (96) depends on the sign of γ^2. There are three separate cases. If γ^2 is positive, the solution is

$$Y = C \cosh(\gamma y) + D \sinh \gamma y)$$

If $\gamma^2 = 0$, the solution is

$$Y = Ey + F$$

If γ^2 is negative, the solution is

$$Y = G \sin(\gamma y) + H \cos(\gamma y)$$

The last is the only solution which is compatible with a resonant condition. If we assume that $Y = 0$ at the center of the bar where $y = 0$ and is a maximum at the edges where $y = \pm \frac{1}{2} w$, then $H = 0$ and $\gamma = (m\pi/w)$.

Substituting the value for ζ and γ in Eq. (97) and solving for k^2 we obtain

$$k^2 = \frac{\omega^2}{\nu^2} - \pi^2 \left(\frac{n^2}{e^2} + \frac{m^2}{w^2} \right) \tag{98}$$

Now $k^2 = (2\pi/\lambda_g)$,^{2} where λ_g is the length of the wave propagated in the z direction in the plate, and $(\omega/v)^2 = (2\pi/\lambda)^2$, where λ is the wavelength in an infinite plate, both waves being excited by the same frequency.

Hence we may write

$$\left(\frac{1}{\lambda_g} \right)^2 = \left(\frac{1}{\lambda} \right)^2 - \frac{1}{4} \left(\frac{n^2}{e^2} + \frac{m^2}{w^2} \right) \tag{99}$$

from which it is easily seen that λ_g is greater than λ unless $m = n = 0$ or unless both e and w are infinite.

Now k^2 in Eq. (98) must be positive or no wave exists. If $k^2 = 0$, the wavelength λ_g is infinite; and if k^2 is negative, the wavelength is imaginary. The wavelength λ for which λ_g is infinite is called the *cutoff wavelength.* It is given by

$$\lambda_c = \frac{2}{(n^2/e^2 + m^2/w^2)^{1/2}} \tag{100}$$

Any actual wave propagated in the rectangular plate must have a wavelength (in an infinite plate) less than λ_c. The frequency $f_c = v/\lambda_c$ is called the *cutoff frequency*. It is the lowest frequency which can be propagated. Waves excited by frequencies lower than f_c cannot be propagated and are attenuated exponentially around the source.

Those familiar with the propagation of electromagnetic waves in wave guides will immediately note the similarity between the behavior of waves traveling in wave guides and waves traveling in solids. The cutoff phenomenon is commonly used in wave guides to eliminate undesired modes and frequencies.

The phenomenon in actual crystalline solids is, of course, much more complicated than that of the hypothetical case developed here. Not only have we ignored the effects of anisotropy but we have also neglected all cross-coupling effects. But the neglect of both effects may not be as serious as might at first be thought. The anisotropy effects in quartz are not large. For example, the three elastic constants involved – c'_{66}, c'_{55}, and c'_{44} – have values 29.3, 69.8, and 41.4×10^9 N/m^2 for the AT-cut orientation. The velocities of the shear waves depending on these constants are in the ratio of the square roots of the numbers or approximately 5:8:6.

The effects of coupling to other modes, through c'_{56}, c'_{12}, etc., have little effect unless the resonant frequencies of the coupled modes are within a few tenths of a percent of each other.

Let us now consider an AT-cut plate of indefinite lateral extent having a thickness e so that its fundamental resonant frequency is given by

$$f_S = \frac{1}{2e}\left(\frac{c'_{66}}{\rho}\right)^{1/2}$$

The area on which the electrodes are plated is designated region E and the surrounding unplated area region S. The mass added by the electrodes makes the effective thickness of region E equal to e'. The frequency of region E is therefore

$$f_E = \frac{1}{2e'} \left(\frac{c'_{66}}{\rho} \right)^{1/2}$$

By means of a signal generator of variable frequency the region under the electrodes can be set into resonance at the frequency f_E. The resulting mode of motion may be described as a plane wave traveling in the thickness direction of the plate. This mode is called the fundamental, or first-harmonic, mode.

As the frequency of the generator is increased other discrete resonances are observed. These are the inharmonic overtone modes, commonly called spurious modes, or simply spurs. They are also associated with standing waves in region E. However, these waves are not propagated in the thickness direction but in directions having components in the plane of the blank. Resonance occurs when these waves are reflected at the discontinuity between regions E and S. The frequencies of these inharmonic modes depend on the dimensions of the plated region and, of course, the thickness of the plate. When the plate is excited at one of the inharmonic overtone frequencies, standing waves may be observed in region E. As stated before, the condition for standing waves to exist is that *reflection must occur at the boundary between regions E and S.*

It is characteristic of any vibrating system that only those modes having frequencies above a certain minimum or cutoff frequency can occur because standing waves of lower frequency cannot exist in the system. The cutoff frequency for region E is f_E and that for region S is f_S. Waves having frequencies higher than f_E and lower than f_S can exist in region E but not in region S, where only waves of frequency equal to or greater than f_S can exist.

Waves in region E having frequencies between f_E and f_S are totally reflected at the boundary between regions E and S. If these waves are excited by the correct frequencies, standing waves may be generated, resulting in spurs. Waves generated by frequencies greater than f_S are not reflected and travel outward toward the edge of the blank.

Suppose that rectangular electrodes having dimensions $w \times 1$ are deposited on region E. The eigenfrequencies of a rectangular (isotropic) plate with perfectly reflecting edges are, from Eq. (37),

$$f_{nmp} = \frac{v}{2}\left[\left(\frac{n}{e}\right)^2 + \left(\frac{m}{w}\right)^2 + \left(\frac{p}{1}\right)^2\right]^{\frac{1}{2}} \qquad \begin{array}{l} n = 1, 3, 5, \ldots \\ m = 1, 2, 3, \ldots \\ p = 1, 2, 3, \ldots \end{array}$$

If w and 1 are each large compared with e, we may write,

$$f_{nmp} = f_{n11}\left[1 + \frac{1}{2}\left(\frac{me}{nw}\right)^2 + \frac{1}{2}\left(\frac{pe}{nl}\right)^2\right]$$

where $f_{n11} = (nv/2e)$ is the frequency of harmonic mode of order n.

The condition for a mode of frequency f_{nmp} *not* to be excited is that $nf_s < f_{nmp}$, which requires that

$$nf_s < f_{n11}\left[1 + \frac{1}{2}\left(\frac{me}{nw}\right)^2 + \frac{1}{2}\left(\frac{pe}{nl}\right)^2\right]$$

and

$$nf_s - f_{n11} < f_{n11} \cdot \left(\frac{1}{2n^2}\right)\left[\left(\frac{me}{w}\right)^2 + \left(\frac{pe}{l}\right)^2\right] \tag{101}$$

Equation (101) gives the plate-back required to suppress the *nmp* mode and all inharmonic modes of higher frequency.

As an illustration of the principle we consider a 10.000 MHz AT-cut plate having square electrodes with dimensions $w = l = 0.300$ cm (0.118 in). The thickness $e = 0.0169$ cm (0.00655 in). The unit is operated on its first-harmonic mode $n = 1$. The first harmonic mode is the 112 (or 121) mode and therefore the plate back must be *less than*

$$10.000 \times 10^6 \times \frac{1}{2}\left[\left(2\ \frac{0.0169}{0.300}\right)^2 + \left(1\ \frac{0.0169}{0.300}\right)^2\right] = 79 \text{ kHz}$$

In order to fabricate such a device the plate is first lapped or etched to a frequency somewhat *below* 10.079 MHz. The electrodes are then applied reducing the frequency to 10.000 MHz. The frequencies of all inharmonic overtones are higher than 10.079 MHz and are therefore not excited, since waves of this and higher frequencies are *not* reflected at the boundary between regions E and S.

To review: Spurious modes can exist in the E region only if waves are reflected at the boundary between E and S. If the frequency of every *potential* mode in E is greater than the cutoff frequency in S, then no wave is reflected and no spurious mode can exist.

As another example we consider the same blank to be supplied with circular electrodes having a diameter of 0.300 cm. The eigenfrequencies of a circular plate with a clamped boundary are found, from Eq. (42), to be

$$f_{nmk} = f_{n01}\left[1 + 2\left(\frac{\chi_{mk}\, e}{n\pi\, d}\right)^2\right] \qquad \begin{array}{l} n = 1, 3, 5, \ldots \\ m = 0, 1, 2, 3, \ldots \\ k = 1, 2, 3, \ldots \end{array}$$

where χ_{mk} is the kth root of Bessel's function of order m. The frequency of the nth harmonic mode is f_{n01}.

The condition that a mode with frequency f_{nmk} *not be excited* is that $nf_s < f_{nmk}$ and therefore that

$$nf_s < f_{n01}\left[1 + \frac{M_{mk}}{n^2}\left(\frac{e}{d}\right)^2\right] \qquad \text{where } M_{mk} = 2\left(\frac{\chi_{mk}}{\pi}\right)^2$$

or that

$$nf_s - f_{n01} < f_{n01}\,\frac{M_{mk}}{n^2}\left(\frac{e}{d}\right)^2 \tag{102}$$

The plate-back required to suppress the first, or $n11$, mode, and all higher modes, can now be computed. The value of $\chi_{11} = 3.832$. Therefore the plate-back must *be less than*

$$10.000 \times 10^6\ \frac{2.98}{n^2}\left(\frac{0.0169}{0.300}\right)^2 = \frac{95}{n^2}\ \text{kHz}$$

Thus, if the resonator is to be used at 10 MHz with all spurs suppressed, the maximum plate-back is 95 kHz. But if the resonator were to be used at its third harmonic at 30 MHz, the maximum plate-back would be only 10 kHz. This might not provide adequate electric conductance.

It is obvious that a tradeoff exists between the thickness/diameter ratio and the plate-back. The allowable plate-back may be increased by reducing the electrode size, but this influences the values of the parameters of the equivalent circuit; in particular, it reduces the value of the motional capacitance C. One solution to this problem is to use the method of "dot" resonators in which several small isolated resonators, all having the same frequency, are operated in parallel.

In order to suppress all inharmonic overtone modes in a resonator with square electrodes we have, from Eq. (101),

$$\text{Plate-back} < f_s \frac{2.5}{n^2}\left(\frac{e}{w}\right)^2 \qquad w = 1 \tag{103}$$

Similarly, in order to suppress all inharmonic overtone modes in a resonator with circular electrodes we have, from Eq. (102),

$$\text{Plate-back} < f_s \frac{2.98}{n^2}\left(\frac{e}{d}\right)^2 \tag{104}$$

In both cases the maximum permissible plate-back is proportional to the frequency of the resonator, inversely proportional to the square of the harmonic order, and proportional to the square of the thickness/lateral dimension ratio.

It should not be expected that the constants 2.5 and 2.98 are exactly correct, because the effects of anisotropy have been ignored and also because a number of approximations have been made in the derivations. But the general forms of Eqs. (103) and (104) do apply and with the constants determined empirically should provide a good starting point for the designer.

It is possible that, in some cases, the first overtone may be absent because of charge cancelation. In such instances, the next overtone is the first one which must be considered and the constants are larger. It must be noted that any convexity of the surfaces of the blank also influences the overtone frequencies.

Finally, some modes which are not suppressed by control of the plate back may often be suppressed by loading the blank at certain points on the unplated area with dots of some cement. The dots are located empirically and experience (and perhaps intuition) helps in determining the correct size and location.

For those readers who are interested in a nonmathematical explanation of energy trapping, we present the following qualitative explanation.

1. An elastic plate having a given set of dimensions, bounded by reflecting edges, has an overtone spectrum of discrete, inharmonically related frequencies.
2. The boundary between the plated and unplated regions of an AT-cut plate acts as a reflecting edge for all acoustic waves having frequencies less than that of the resonant frequency of the unplated area.
3. If the frequency of the unplated area is *less* than that of the overtone of lowest frequency, then the plated area can be excited into resonance *only* at its own harmonic frequency (provided no other reflecting edges are present).

A more complete discussion of the energy-trapping concept is given by Shockley, Curran, and Koneval.[1] If Eq. (24) of this paper is solved for $f_e - f_s$, the result is the same as Eqs. (103) and (104) except for the values of the constants. Those interested in a more complete discussion of this subject should consult the original article.

---

[1] *J. Acoust. Soc. Amer.*, 41:985–993 (1967).

11
X – RAYS AND X – RAY ORIENTATION

INTRODUCTION

We have seen in Chap. 8 that the frequency-temperature characteristics of a vibrating quartz piezoid depend on its orientation with respect to the crystallographic axes. In particular, Eq. (92) shows that the frequency-temperature coefficient T_f for the rotated Y-cut family which includes the AT- and BT-cuts is a function of the angle θ which is the angle between the plane of the quartz blank and the Z-axis. (The angle θ is often called the ZZ' angle.) Figures 8.3 to 8.5 show how T_f is affected by changes in the angle θ. For example, from Fig. 8.3 we see that a 100 MHz AT-cut plate with a ZZ' angle of 35°26′ exhibits a frequency change of only ± 12 ppm over a temperature range of –60 to +100°C. The frequency excursion is doubled if the ZZ' angle is increased or decreased by only 4′.

It is apparent that precise control of the angle of orientation is required in order to take advantage of the inherent temperature stability of the quartz crystal resonator. Other factors, including the electrode material and thickness, the dimensions and contour of the blank, and mounting stresses also influence the frequency-temperature characteristics, but the dominant factor is the orientation.

Before the development of x-ray diffraction equipment, quartz piezoids were oriented by reference to the natural faces of the crystal. For example, an AT-cut can be made by sawing a blank, making a dihedral angle of about 3° with respect to a minor rhomb (r) face of the natural crystal; the line of intersection of the two planes is the X-axis. This is readily seen from Fig. 8.2, where it is shown that the r face makes an angle of 38°13′ with the Z-axis and the AT-cut makes an angle of about 35°15′ (the exact angle depending on the application) with the *Z*-axis.

Using crystals with well-developed natural faces, and with skill and care, it is possible to cut blanks with an accuracy of perhaps ±30′. However, in the absence of good natural faces and where greater precision is required, x-ray diffraction methods are necessary. Without x-ray diffraction it would be quite impossible to achieve the precision required in most applications of piezoelectric resonators. It is interesting to note that the first commercial application of x-ray diffraction was made in the quartz crystal industry during the years of World War II.

NATURE AND PRODUCTION OF X-RAYS

X-rays are very short electromagnetic waves. They are produced when electrons traveling at high velocities strike a stationary atom. An x-ray tube is a high vacuum diode having a hot filament or cathode which provides a supply of electrons, and an anode or target upon which the electrons impinge. The x-ray tubes used in the quartz crystal industry always have copper targets. The reason for this will appear later. The anode must be at a high positive potential with respect to the cathode. Some tubes are operated on alternating current. In this case the tube acts as a rectifier and x-rays are produced only during that part of the cycle during which the anode is positive relative to the cathode. Currents of 1 to 50 mA are commonly employed. The anode may be either air- or water-cooled.

The spectrum of x-rays emitted by the target consists of a continuous range of wavelengths (Bremsstrahlung) with the shortest wavelength depending on the peak voltage across the tube and the longest wavelength depending on the material of the window of the tube. In addition to the continuous spectrum, often called *white radiation*, the tube emits intense radiation of discrete wavelengths (provided the voltage is high enough) which depends upon the atoms of which the target is made. The strongest of these characteristic radiations is called the $K\alpha_1$ radiation or wavelength. Very close to it in wavelength and about half as intense is another line called the $K\alpha_2$ radiation. Another pair of lines of somewhat shorter wavelength and about 15 percent the intensity is called the $K\beta_1$ and $K\beta_2$ radiation. The manner in which the intensity of the radiation from a copper target varies with wavelength at two different voltages is shown in Fig. 11.1. At voltages lower than about 8 kV the $K\alpha$ radiation is not excited. Any voltage greater than this threshold voltage may be used to produce the $K\alpha$ radiation, but in order to obtain adequate intensity, voltages of 25 to 50 kV are commonly employed. The wavelength of the characteristic radiation does not depend on the voltage or the current or even the temperature but only on the atoms of which the target is made. It thus provides a fixed standard of length for making measurements.

The wavelengths of the characteristic radiations have been carefully determined and the values for copper are given in Table 11.1.

BRAGG DIFFRACTION

A beam of x-rays incident on a crystal may be scattered coherently if certain conditions are met. This coherent scattering is often called *reflection*, although unlike optical reflection, it can occur only at specific angles. Sometimes the word *diffraction* is used, but this term is not strictly applicable either, at least in the optical sense. We will use the term *reflection* with the understanding that we refer to the phenomenon of coherent scattering from the atomic planes of a crystal.

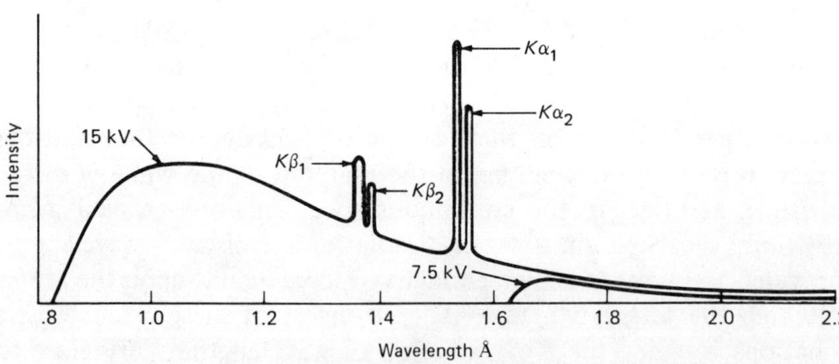

Fig. 11.1. Variation of the intensity of the radiation from a copper target at 7.5 and 15 kV. The intensity is not shown to scale; the characteristic intensities are relatively much greater.

Table 11.1. Characteristic Wavelengths for Copper.

$K\alpha_1$ = 1.537395 Å	$K\beta_1$ = 1.38935 Å
$K\alpha_2$ = 1.541232	$K\beta_2$ = 1.37824

Note: x-ray wavelengths are often given in XU. One Å is equal to 1×10^{-8} cm and 1000 XU = 1.00202 Å. The difference is not important for our purposes.

The condition for coherent scattering or reflection of x-rays from a crystal is known as the Bragg relation, which is written

$$n\lambda = 2d \sin \theta_B \qquad n = 1, 2, 3, \ldots \tag{105}$$

where λ = wavelength of the x-rays
d = distance between planes of atoms in the crystal
θ_B = angle (called *Bragg angle*) at which reflection occurs

The integer n is called the order of the reflection. Usually n is not greater than 3.

The origin of the Bragg relation is easily seen from Fig. 11.2. The horizontal rows of dots represent planes of atoms separated by the distance d. A beam of monochromatic x-rays of wavelength λ is incident from the left. In order for this beam to be reflected coherently (diffracted), the path difference $AB - AC$ must be equal to an integral number of wavelengths. But $AB = d/\sin \theta$ and $AC = AB \cos 2\theta$. Therefore

$$n\lambda = AB - AC = \frac{d}{\sin 0} - \frac{d \cos 2\theta}{\sin 0} = 2d \sin \theta$$

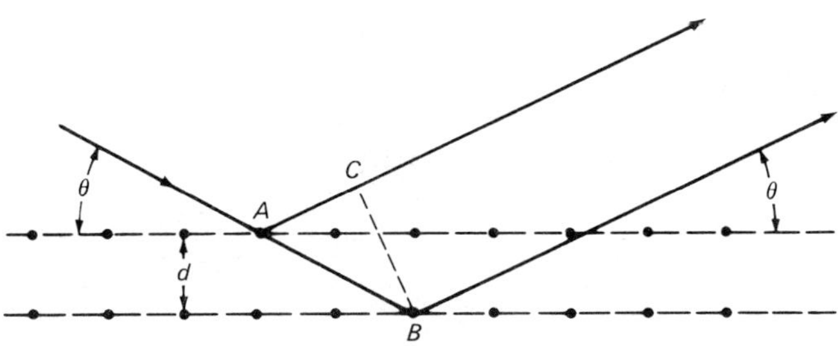

Fig. 11.2. The origin of the Bragg relation.

It must be remembered that x-ray reflection, unlike optical reflection, does not occur from the surface but from planes of atoms which may or may not be parallel to the surface of the specimen. This reflection involves the first few hundred atomic planes. It should also be noted that the angle of reflection, i.e., the Bragg angle θ_B, is always taken to be the angle between the incident beam and the reflecting planes in contrast to the custom in optical work where the angle of incidence is defined to be the angle between the incident beam and the *normal* to the reflecting surface.

It follows from the Bragg relation that if the wavelength of the x-rays is known, then the atomic spacing d can be determined by measuring the angle at which reflection occurs. Conversely, if d is known, then the wavelength of the x-rays can be measured. In the use of x-ray diffraction for crystal orientation, all three quantities λ, θ_B, and d are known and the problem is always to determine the dihedral angle between the plane surface of a quartz piezoid and a particular set of atomic planes. The equipment used in doing so is described in the following sections.

The value of d for any set $(hk.l)$ of atomic planes in quartz can be determined if the two lattice parameters a and c are known. In quartz the values are

$$a = 4.903 \text{ Å}$$

at 25°C

and

$$c = 5.392$$

$$d_{(hk.l)} = \frac{a}{\sqrt{\frac{4}{3}(h^2 + hk + k^2) + (al/c)^2}}$$

It is also possible to compute the Bragg angle for any set of planes from the relationship

$$\sin^2 \theta_B = \frac{\lambda^2}{4}\left(\frac{4}{3}\;\frac{h^2 + hk + k^2}{a^2} + \frac{l^2}{c^2}\right)$$

The Bragg angles depend slightly on the temperature of the crystal through the variation of d with temperature. Most measurements are made at temperatures near room temperature (25°C) and the variation of θ_B with temperature can usually be neglected in quartz orientation.

Table 11.2 shows the Bragg angles and the B-M indices for a few of the atomic planes commonly used in quartz orientation. It is assumed that Cu $K\alpha_1$ radiation is used and that the temperature is 25°C. The values are considered to be the best values available at the present time. Work is continuing to refine the values, but this is of little consequence in quartz crystal orientation since it is not the absolute value of θ_B which is required but, as stated before, the dihedral angle between the atomic planes and the surface of the quartz plate or piezoid.

Equation (105) requires that $n\lambda/2d \leqslant 1$. It is usually not convenient to use Bragg angles near 0 or 90° because in the first case the incident beam enters the detector and in the second the detector blocks the incident beam. It is usually desirable to use first-order reflections because the intensity is greater. If $0.3d < \lambda < 2d$, then $10° < \theta_B < 80°$, which is a convenient range in which to work. It is thus apparent that the wavelength must be comparable to the spacing. This requirement is nicely satisfied by the Cu $K\alpha_1$ line and for this reason the x-ray tubes used in the crystal industry usually have copper targets. We often specify the value of the Bragg angle without mentioning that the use of the Cu $K\alpha_1$ radiation is understood.

CARE IN THE USE OF X-RAYS

Although the intensity of the x radiation used in crystal orientation is low and the equipment is well shielded, it must never be forgotten that exposure to x-rays at any level should be avoided. Operators

Table 11.2. The Bragg Angles and B-M Indices of the Atomic Planes Used in Quartz Crystal Orientation.*

Designation	B-M Indices	Bragg Angle
Y or *m* face	(10.0)	10° 26′
X-planes	$(2\bar{1}.0)$	18° 18′
Z-planes	(00.3)	25° 19′
Used with BT-cut	(20.3)	34° 03′
R face (major rhomb)	$(\bar{1}0.1)$	13° 19′
r face (minor rhomb)	(10.1)	13° 19′
Used with SC- and IT-cuts	(12.2)	33° 53′

*It is assumed that Cu $K\alpha_1$ radiation is used.

should be required to keep their hands out of and away from the direct beam. Appropriate radiation monitoring devices (dosimeters) should be used by all personnel working with x-ray diffraction equipment. Periodic inspections should be made to ensure that all shields are in place and that no leakage of radiation is occurring. Finally, permanent records of the radiation received, even if it is zero, should be kept on file.

SINGLE-CRYSTAL X-RAY DIFFRACTION SYSTEM

The simplest type of x-ray diffraction equipment consists of an x-ray tube and power supply, a set of collimating slits, a goniometer and a suitable chuck for holding the crystal blank, a nickel filter, and a detector. The arrangement of the apparatus is shown in Fig. 11.3.

The collimating slits S_1 and S_2 are used to define a parallel beam of x-rays which passes through the axis of rotation O of the goniometer. The nickel filter, having a thickness of the order of 0.01 mm, absorbs the $K\beta$ radiation and passes the more intense $K\alpha$ radiation. The quartz plate or blank is held in place by a vacuum chuck system consisting of three sapphire beads so disposed that the reflecting surface of the blank lies in the goniometer axis. The orientation errors which result from failure of the incident beam to intersect the goniometer axis are discussed at the end of this chapter.

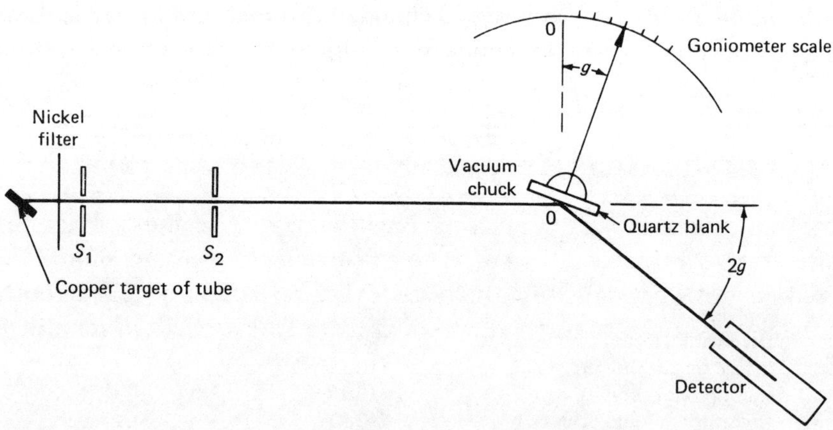

Fig. 11.3. The single-crystal diffraction system.

The goniometer should be capable of reading angles with an accuracy of 1′ of arc or better. It rotates the quartz blank about its center *O*, thereby varying the angle which the plane of the blank makes with the incident beam. When the angle between the atomic planes and the incident beam is equal to the Bragg angle θ_B, the beam is reflected and reaches the detector. Various types of detectors are used. High-intensity beams are usually detected by means of ionization chambers. Geiger (GM) counters with suitable integrating circuits are used with low-intensity beams. Solid-state detectors are also used.

The precision available with the apparatus just described is limited by the slit width and the natural line width of the $K\alpha$ radiation. It is usually not possible to resolve the $K\alpha_1$ and $K\alpha_2$ radiation with such apparatus. These wavelengths are reflected at angles differing typically by 2 to 5′ depending on θ_B. The minimum slit width is determined by intensity requirements. Because of these limitations orientation accuracy with the single crystal system is limited to about ± 5′ and in practical use even this accuracy is scarcely achievable.

DOUBLE-CRYSTAL X-RAY DIFFRACTION SYSTEM

It has become almost universal practice in the quartz crystal industry to use the double-crystal x-ray diffraction system which is illustrated in Fig. 11.4. The radiation from the x-ray tube is directed onto the surface of crystal *A*, which is called the collimating (or monochromator) crystal. Only that radiation which strikes crystal *A* at the correct Bragg angle is reflected, so that collimating slits and the nickel filter may be omitted. Crystal *A* is carefully prepared and permanently mounted to reflect the Cu $K\alpha_1$ radiation. The reflected radiation strikes the surface of crystal *B*, which is supported by the vacuum chuck in such a position that the axis *O* of the goniometer intersects the incident beam. The same reflecting planes should be used in both crystals. In this case Bragg reflection occurs at both crystals when the two sets of atomic planes are parallel to each other and to the goniometer axis.

Two conditions must be satisfied in order to take full advantage of the double-crystal system. First, the same reflecting planes must be used in both crystals and second, the atomic reflecting planes must

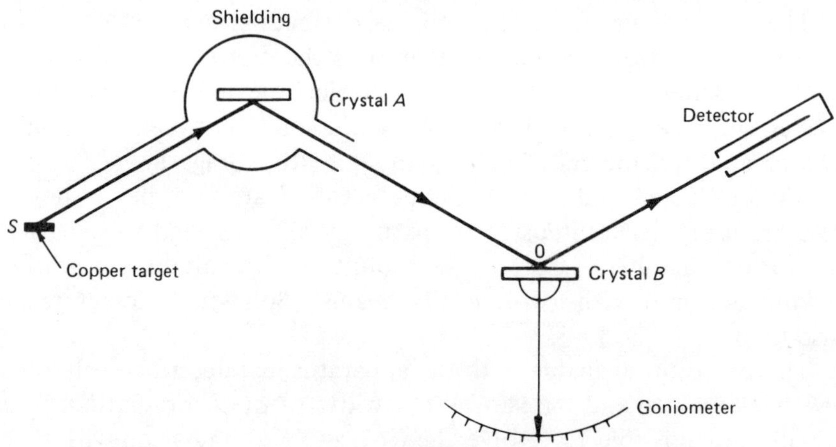

Fig. 11.4. Double-crystal x-ray diffraction system.

be parallel. In addition, it is desirable that the collimating crystal be nearer the x-ray tube. The positions of the fixed and moveable crystals may be interchanged without loss of accuracy but at some sacrifice of convenience and flexibility.

The advantages of the double-crystal system can be seen by considering Fig. 11.5. Here S is the focal spot on the target of the x-ray tube. The width of S is typically 2 to 3 mm. Each point on the focal spot emits x-rays of all wavelengths (see Fig. 11.1) in all directions. We consider the radiation in a band of wavelengths between λ_1 and λ_2 which may be the $K\alpha_1$ and $K\alpha_2$ or the $K\alpha$ and $K\beta$ radiations. Radiation of wavelength λ_1 is reflected from crystal A if and only if the Bragg relation is satisfied. Hence, radiation of wavelength λ_1 from every point on S is reflected at angle θ_{B_1} from *some* point on

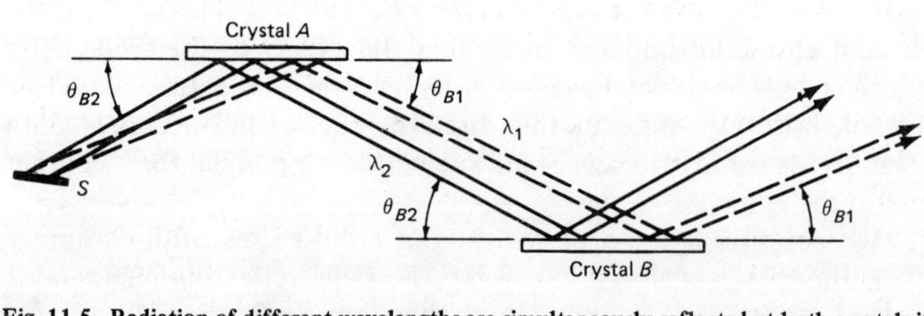

Fig. 11.5. Radiation of different wavelengths are simultaneously reflected at both crystals if they are in the parallel position.

crystal A. The same can be said for radiation of wavelength λ_2. The beams of wavelength λ_1 and λ_2, reflected from crystal A, are not parallel. Their angular separation can be calculated in the following way.

From the Bragg relation:

$$n\lambda = 2d \sin \theta_B$$

$$n d\lambda = 2d \cos \theta_B \; d\theta_B$$

Dividing the second equation by the first,

$$\frac{d\lambda}{\lambda} = \cot \theta_B \; d\theta_B$$

or

$$d\theta_B = \tan \theta_B \frac{d\lambda}{\lambda} \quad \text{radians} \qquad (106)$$

If we let $d\lambda$ be the difference between the wavelengths of the $K\alpha_1$ and $K\alpha_2$ radiations and take $\theta_B = 13°19'$ corresponding to reflection from the (10.1) planes, we obtain

$$d\theta_B = \theta_{B_1} - \theta_{B_2} = 2.0'$$

After reflection from crystal A the two beams strike crystal B, which has the same diffraction constant (atomic spacing) as crystal A. If and only if the atomic planes of the two crystals are parallel, *both* beams are simultaneously reflected from crystal B. Thus the angular measurement compares the alignment of the atomic planes in the two crystals and the peak reading occurs when the two sets of planes are exactly parallel. The wavelength of the radiation need not be known.

If the lattice spacing of crystal B is not the same as that of crystal A or if the atomic planes of the two are not exactly parallel, then λ_1 may be reflected at crystal B but not at angle θ_{B_1}. Similarly λ_2 may be reflected but not at angle θ_{B_2}. Hence, in general, a "double peak" results, which is why the reflecting planes in the two crystals should be identical and parallel to the axis of the goniometer. Unless crystals A and B satisfy both these conditions, the advantages of the double-crystal system are not fully utilized.

Double-crystal reflection can be obtained with the crystals in the antiparallel position shown in Fig. 11.6, but in this position if

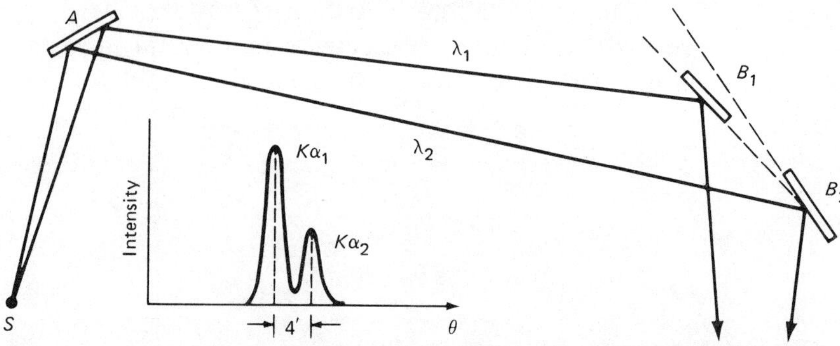

Fig. 11.6. The antiparallel position. Crystal *A* is fixed. B_1 is the position of crystal *B* to reflect λ_1 and B_2 is its position to reflect λ_2. The change in the angular position is given by $d\theta = 2 \tan \theta_B \, d\lambda/\lambda$. For Cu K_α radiation and the (10.1) planes $d\theta = 4.0'$. The insert shows the resulting double peak.

crystal *B* is set to reflect λ_1 it cannot reflect λ_2, and vice versa. For example, if crystals *A* and *B* employ the (10.1) planes and are set to reflect the $K\alpha_1$ radiation, one of the two crystals must be rotated by 4.0′ in order to reflect the $K\alpha_2$ radiation, thus resulting in the double peak mentioned above. Since the presence of the double peak may (and usually does) result in confusion, the antiparallel position should be avoided in designing a double-crystal diffraction system.

The intensity of the reflected radiation received by the detector is usually greater in the double-crystal system than in the single-crystal system despite the two reflections. This is because the radiation from the entire focal spot may be utilized, since no narrow collimating slits are required and also because a wider range of wavelengths is used. The width of the "rocking curve" (a plot of the intensity of the radiation received by the detector vs. the angular position of the crystal) depends not on the range of wavelengths, including the natural line width, but only on the broadening due to the crystal itself. In other words, the wavelength dispersion in the double-crystal system is zero, but the dispersion due to crystalline imperfections is not. In the single-crystal system both types of dispersion are present.

The most important cause of crystal dispersion is that of misoriented crystallites on the surface of the crystal. Any quartz surface which has been lapped or polished mechanically always has misoriented crystallites on the surface. The degree of misorientation can be minimized by careful lapping and polishing, but etching with HF or NH_4F_2 deeply enough to remove the misoriented quartz is necessary to reduce the rocking curve to its limit of about 1′ of arc. The width cannot be reduced below this value for reasons which we cannot pursue here. When greater precision is required, it is necessary to find the center of the rocking curve by some means.

PREPARATION OF CRYSTAL STANDARDS AND COLLIMATING CRYSTALS

It is convenient and almost necessary in adjusting and aligning an x-ray system to have a set of "standard crystals." These are crystal blanks so cut that they have a major surface which is parallel to a given set of atomic planes. The surface of the blank should be polished and the angle adjusted until the blank can be rotated about an axis perpendicular to its reflecting surface without a change in the reflecting angle. This ensures that the atomic planes are parallel to the surface. Flatness should be checked with an optical flat. The blank should then be etched to remove all misoriented material as indicated by the width of the rocking curve. It will be noted that the intensity of the reflected beam decreases with etching.

Natural *r* or *R* faces are approximately parallel to the (10.1) planes and therefore provide a good start for preparing a (10.1) standard. But angle correction, polishing, and etching are still necessary to make a good standard.

Collimating crystals for use in double-crystal systems should be prepared in the same manner. Both standard and collimating crystals should be thick enough to prevent bending as they are mounted or used.

ANGLE MEASUREMENTS BY X-RAY DIFFRACTION

In the use of x-ray diffraction for crystal orientation, the object of all measurements is to determine the angle between a plane surface on the piezoid and a specified set of atomic planes. As a first example, suppose that we desire to make and check a *Y*-cut blank. The major surfaces of a *Y*-cut are normal to the *Y*-axis or parallel to the (10.0) planes which have a Bragg angle of 10°26′. If the surface of the plate is exactly parallel to the (10.0) planes, then reflection of the $K\alpha_1$ radiation occurs when the incident beam makes an angle of 10°26′ with the surface of the plate and the plate can be rotated about an axis normal to its surface with no change in the goniometer reading. This case is shown in Fig. 11.7*a*.

In Fig. 11.7*b* we have the case in which the atomic planes make a dihedral angle δ with the surface of the plate; the line of intersection being parallel to the axis of the goniometer. Now the angle of incidence with the surface (not the atomic planes) is $\theta_B + \delta$ and the angle of reflection is $\theta_B - \delta$. In actual practice it is the angle δ which is required.

Now if the blank is rotated by 180° about the normal to its surface, the angle of incidence becomes $\theta_B - \delta$ and the angle of reflection $\theta_B + \delta$. The angle δ can be determined in either of two ways. We may use a standard blank in which $\delta = 0$, as shown in Fig. 11.7*a*, to determine θ_B and take the difference between this and the reading obtained from the blank, as shown in Fig. 11.7*b*. Alternatively we may take one reading, rotate the blank in its plane by 180°, and take a second reading. One-half the sum of the two readings is θ_B and one-half the difference is δ. However both readings must be made with the line of intersection of the atomic planes and the blank surface parallel to the goniometer axis.

The method of Fig. 11.7*b* is used to determine the *ZZ′*-angle of the AT-cut. (The *ZZ′*-angle is the angle between the plane of the blank and the *Z*-axis of the crystal.) If the AT-cut is properly made

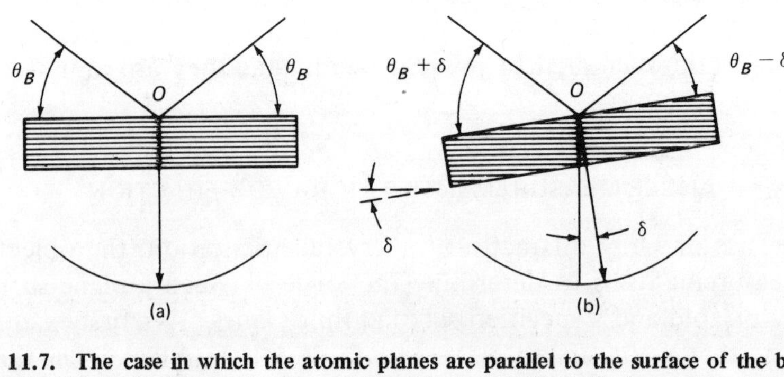

Fig. 11.7. The case in which the atomic planes are parallel to the surface of the blank is shown in (*a*). In (*b*) the atomic planes make a dihedral angle δ with the surface of the blank.

the angle δ between the surface of the blank and the (10.1) planes is about 3° (depending on the application) and the line of intersection of the planes is parallel to the X-axis. The quartz blank is placed on the chuck with the X-axis parallel to the axis of the goniometer. In the reflecting position the goniometer reading $g = \theta_B \pm \delta$, the sign depending on whether the X-axis points up or down. The ZZ'-angle is then either $38°13' + \delta$ or $38°13' - \delta$. The correct position is, of course, $38°13' - \delta$. Care must be taken to avoid the mistake of cutting the blank at $38°13' + \delta$, because the resulting blank will have a large negative temperature coefficient. This mistake may not be observed with a single x-ray measurement.

ANGLE MEASUREMENTS WHEN THE ATOMIC PLANES DO NOT CONTAIN THE GONIOMETER AXIS

The first case which we examine is that in which the line of intersection of the atomic planes and the blank surface is perpendicular to the goniometer axis, as shown in Fig. 11.8. The plane of the surface of the blank is parallel to OO, the goniometer axis, but the atomic planes make an angle δ with it. The incident beam and the

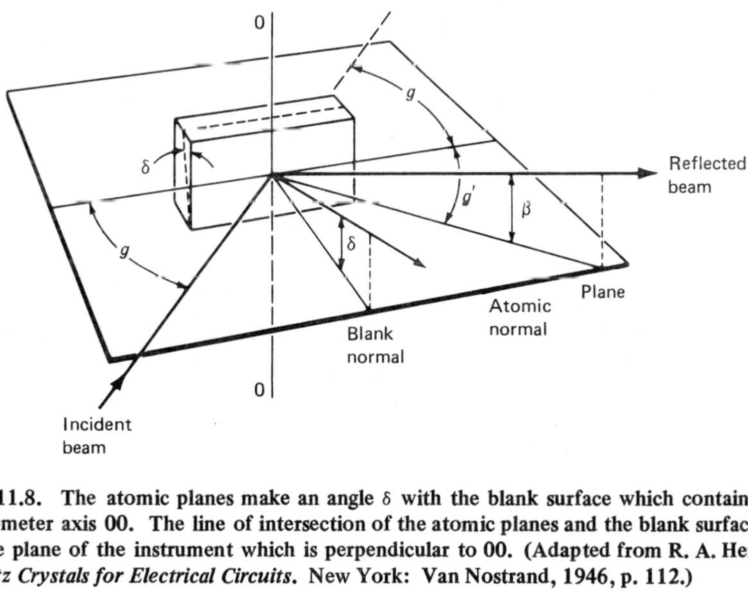

Fig. 11.8. The atomic planes make an angle δ with the blank surface which contains the goniometer axis 00. The line of intersection of the atomic planes and the blank surface lies in the plane of the instrument which is perpendicular to 00. (Adapted from R. A. Heising, *Quartz Crystals for Electrical Circuits*. New York: Van Nostrand, 1946, p. 112.)

normal to the blank lie in the plane of the instrument, but the normal to the atomic planes and the reflected beam do not do so.

The incident beam, the normal to the atomic planes, and the reflected beam lie in a plane which is called the *reflection plane.* In this plane the angle of incidence is θ_B and the detector setting is $2\theta_B$. However, the angles are not measured in this plane but in the plane of the instrument. Therefore, the goniometer reading g and the detector setting $g + g'$ are the projections of the angles θ_B and $2\theta_B$ on the plane of the instrument. The relationships between these quantities are[1]

$$\sin g = \frac{\sin \theta_B}{\cos \delta} \qquad 0 \leqslant \delta < 90° - \theta_B \tag{107}$$

and

$$\tan g' = \tan g \cos 2\delta \tag{108}$$

For example, if an AT-cut blank in which $\delta = 2°58'$ is placed on the goniometer chuck with the X-axis normal to OO (Fig. 11.8), reflection is obtained from the (10.1) planes for which $\theta_B = 13°19.0'$ at

$$g = 13°20'$$

and

$$g' = 13°16'$$

making the detector angle $g + g' = 26°36'$ instead of $26°38'$. Thus the "apparent Bragg angle" g is larger than the actual Bragg angle by about $1'$. Usually this difference is negligible. If, however, the angle δ were larger, say $10°$, then from Eq. (107) $g = 13°32'$, thereby requiring a correction of $13'$. From Eq. (108) the value of $g' = 12°45'$, thereby making the detector position $g + g'$ have the value of $26°17'$.

If δ is large the reflected beam may be directed above or below the detector. The angle β by which the reflected beam is directed above or below the goniometer plane is given by

$$\sin \beta = 2 \sin \delta \sin \theta_B = \sin 2\delta \sin g \, . \tag{109}$$

In the last example where $\theta_B = 13°19'$ and $\delta = 10°$ we have $\beta = 4°35'$.

---

[1] The derivations of Eqs. (107) through (112) are given at the end of this chapter.

It is therefore apparent that the atomic planes used in checking the orientation of a given blank should make a small dihedral angle with the surface of the blank; otherwise corrections may be required for the errors introduced and relocation of the detector may be necessary.

THE EFFECT OF ROTATION IN THE PLANE OF THE BLANK

In measuring the ZZ' angle of the AT-cut blank it has been assumed that the X-axis lies in the plane of the blank and is parallel to the goniometer axis. Likewise, in measuring the XX' angle, the X-axis was assumed to be perpendicular to the goniometer axis. Errors are introduced if these conditions are not satisfied. We now investigate this source of error.

Let ψ be the angle of rotation of the blank about a normal to its surface. We assume that the line of intersection of the atomic planes and the blank surface is the X-axis and that $\psi = 0$ when the X-axis lies in the goniometer plane. Equation (110*a*) relates the angular position ψ with the goniometer reading g.

$$\cos^2 \theta_B = \sin^2 \delta \cos^2 \psi + (\cos g \cos \delta - \sin g \sin \delta \sin \psi)^2 \qquad (110a)$$

Equation (110*a*) can be solved only by approximation methods, which is rather tedious. Equation (110*b*) gives values of g to a very high degree of approximation for all reasonable values of δ and is easily soluble for g.

$$\sin (g + \delta \sin \psi) \doteq \frac{\sin \theta_B}{\cos (\delta \cos \psi)} \qquad (110b)$$

Both equations reduce to $g = \theta_B$ if $\delta = 0$ for all values of ψ; to $g = \theta_B - \delta$ if $\psi = 90°$ and to $g = \theta_B + \delta$ if $\psi = 270°$ and to Eq. (107) if $\psi = 0$.

The extreme values of g occur at the positions $\psi = 90°$ and $270°$, where $g = \theta_B \pm \delta$ thereby providing a method for determining the position $\psi = 0$ when the X-axis lies in the goniometer plane. However $g \neq \theta_B$ at $\psi = 0°$ or $180°$ but at angles slightly above $0°$ and below $180°$. We have already seen that in the case of the AT-cut, if

$\psi = 0$, the difference between g and θ_B is about 1′. Figure 11.9 shows the manner in which g varies with ψ.

It is important to know how much error is introduced into the measurement of the ZZ' angle by errors in the angle ψ. As an example we take $\psi = 80°$, $\delta = 2°58'$, and $\theta_B = 13°19'$. Substituting these values in Eq. (110b) we find $g = 10°23.7'$. When $\psi = 90°$, $g = (\theta_B - \delta) = 10°21'$. Therefore an error of 2.7′ is introduced by a rotation of the blank by 10° about its normal. A similar calculation shows that a rotation of ± 5° introduces an error of only 0.7′.

CHECKING DOUBLY ROTATED BLANKS BY X-RAY DIFFRACTION

In the rotated Y-cut family which includes the AT- and BT-cuts as well as several others, the line of intersection of a suitable set of atomic planes and the surface of the blank is the X-axis. However, several cuts are known for which this is not the case; one example being the SC-cut (stress compensated) which is quite important because its frequency is substantially independent of stresses in the plane of the blank. We now show how x-ray diffraction may be used to determine the orientation of blanks such as the SC-cut.

It is convenient to specify the orientation of atomic planes and surfaces of blanks by the angles θ and ϕ in the coordinate system shown in Fig. 11.10. The angle θ is the angle between the Z-axis and the plane of the blank or the atomic plane. The angle ϕ is the angle between the X-axis and the line of intersection of the blank or atomic plane with the XY-plane of the coordinate system.

A blank which can be generated by a single rotation about one of the XYZ-crystallographic axes is called a *singly rotated* blank. The AT- and BT-cuts are examples of singly rotated blanks. If two rota-

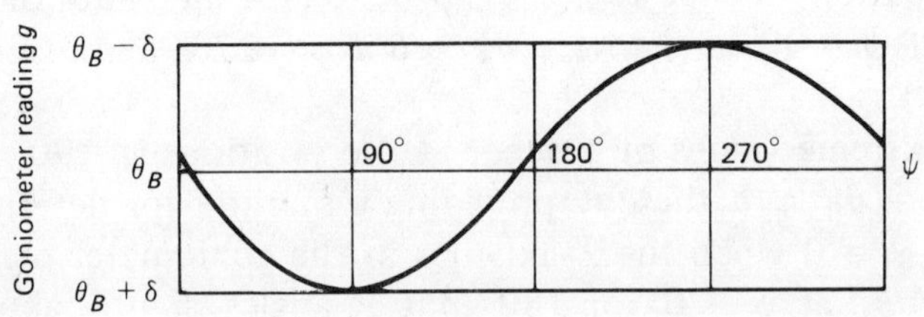

Fig. 11.9. Variation of the goniometer reading with the angle ψ for the AT-cut.

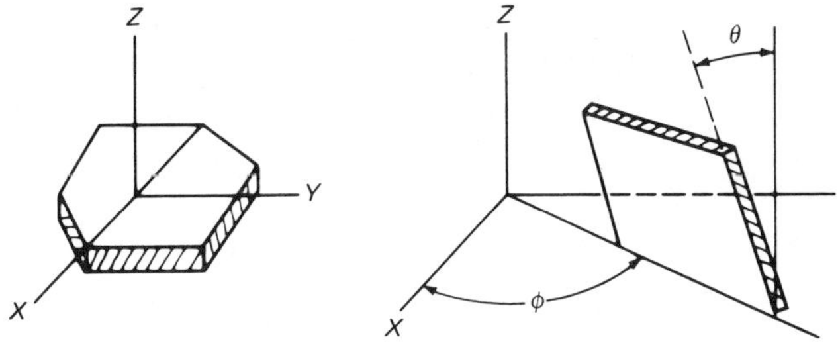

Fig. 11.10. Orientation of a double rotation plane in quartz.

tions are required the blank is said to be *doubly rotated.* Examples are the IT- and SC-cuts.

We have already seen that checking the value of θ for a singly rotated cut is relatively simple. It is only necessary to measure the angle between the surface of the blank and the atomic planes and then to add this angle to the value of θ for the atomic planes being used. Checking for a possible error in the XX' angle, which is really a ϕ angle, is a bit more complicated if the angle between the surface of the blank and the atomic planes is not very small. Such a blank is really a doubly rotated cut.

Table 11.3. Values of θ, ϕ, and θ_B for Some of the Planes Used in Orienting Quartz Piezoids.

Designation	θ	ϕ	Bragg angle θ_B
X-cut	0°	90°	
(11.0)	0°	90°	18° 18′
Y-cut	0°	0°	
(10.0)	0°	0°	10° 26′
AT-cut	35° 15′*	0	
(10.1)	38° 13′	0	13° 19′
BT-cut	49° *	60°	
(20.3)	49° 44′	60°	34° 03′
IT-cut	34° 20′*	19° 06′	
(12.2)	30° 46′	19° 06′	33° 53′
SC-cut	34° 07′*	21° 56′*	

*These angles are subject to small variations depending on the particular applications.

Checking the orientation of a doubly rotated cut is complicated if it happens that neither θ nor ϕ is the same for the blank and any set of atomic planes. Nevertheless, the orientation can be checked with arbitrary accuracy by a method to be described. Later we will use the SC-cut to illustrate the process.

We assume that the blank has been cut in such a way that one edge, which we will call the "reference edge," lies in the XY-plane. This presents no problem, since the XY-plane can be cut on the original crystal, checked by x-ray diffraction, and corrected if necessary. In fact, the manufacturers of cultured quartz supply "prelumbered" bars with accurately oriented XY- and YZ-surfaces. These surfaces can be painted and later identified on the blanks cut from the crystal.

We begin by defining two difference angles (see Fig. 11.13).

$$D_\theta = \theta_{\text{blank}} - \theta_{(hk.l)}$$

and

$$D_\phi = \phi_{\text{blank}} - \phi_{(hk.l)}$$

We assume that D_θ and D_ϕ are small, i.e., that a set of atomic planes exists near the required cut. Values of D_θ and D_ϕ as large as 10 or 15° are tolerable although smaller values are certainly desirable.

The blank is placed on the x-ray chuck with the reference edge vertical (parallel to the goniometer axis). The angle at which reflection is observed is read on the goniometer and designated g_θ. The blank is then placed on the chuck with the reference edge horizontal. The angle read is called g_ϕ.

Now if both D_θ and D_ϕ were zero, we should have $g_\theta = g_\phi = \theta_B$. If either D_θ or D_ϕ were zero, the value of the other would be

$$D_\theta = g_\theta - \theta_B \qquad (D_\phi = 0)$$

$$D_\phi = g_\phi - \theta_B \qquad (D_\theta = 0)$$

If neither D_θ nor D_ϕ is zero or negligibly small, then we must write

$$\sin(g_\theta - D_\theta) = \frac{\sin\theta_B}{\cos D_\phi} \tag{111}$$

and

$$\sin(g_\phi - D_\phi) = \frac{\sin\theta_B}{\cos D_\theta} \tag{112}$$

The proofs of Eqs. (111) and (112) are given at the end of this chapter.

We can find the values of D_θ and D_ϕ satisfying these two equations by a process of iteration which converges very rapidly because the value of one of the quantities has only a second-order effect on the other.

We take for a zeroth approximation

$$_0D_\theta = g_\theta - \theta_B$$

and use this value of D_θ to calculate the first approximation to D_ϕ. Using Eq. (112),

$$\sin (g_\phi - {}_1D_\phi) = \frac{\sin \theta_B}{\cos {}_0D_\theta}$$

Then using the value of $_1D_\phi$ thus found, we substitute in Eq. (111) to find $_1D_\theta$. Continuing the process for two or three steps enables one to find the values of D_θ and D_ϕ. The process converges so rapidly that even for values of D_θ and D_ϕ as large as 10°, the results are accurate to 0.001° after three iterations.

The values of θ and ϕ for the blank are then given by

$$\theta_{\text{blank}} = \theta_{(hk.l)} + D_\theta \qquad \text{and} \qquad \phi_{\text{blank}} = \phi_{(hk.l)} + D_\phi \tag{113}$$

The procedure is a bit tedious even with a hand calculator. But with a programmable calculator or a computer it is only necessary to supply the values of g_θ and g_ϕ to obtain the values of the orientation angles θ and ϕ of the blank.

We will now illustrate the process by computing the values of θ and ϕ for an SC-cut blank. The example chosen is a rather extreme case for purposes of illustration; ordinarily the initial cut would be made closer to the required orientation. The (12.2) planes are used. Note that the collimating crystal should also use the same planes to avoid a double peak.

For purposes of illustration we assume goniometer readings of

$$g_\theta = 47.000° \qquad \text{and} \qquad g_\phi = 29.000°$$

(If the SC-cut were intended to have angles θ = 34°07′ and ϕ = 21°56′, the values of g_θ and g_ϕ, respectively, should be 37°16′ and

36°48′. Thus in the hypothetical example the blank is cut far from the correct orientation.)

The Bragg angle θ_B for the (12.2) planes is 33.8833°. In the zeroth approximation

$$ {}_0D_\theta = g_\theta - \theta_B = 13.1167° \qquad \text{and} \qquad {}_0D_\phi = g_\phi - \theta_B = -4.8833° $$

We may now use either of these values to calculate a first approximation to the other. Thus using Eq. (111),

$$ \sin\,(g_\theta - {}_1D_\theta) = \frac{\sin\theta_B}{\cos {}_0D_\phi} $$

we have

$$ {}_1D_\theta = 12.9764° $$

This value may now be used with Eq. (112) to compute ${}_1D_\phi$; thus

$$ \sin\,(g_\phi - {}_1D_\phi) = \frac{\sin\theta_B}{\cos {}_1D_\theta} $$

from which

$$ {}_1D_\phi = -5.8977° $$

Continuing in this way we obtain

$$ {}_2D_\theta = 12.9117° \qquad \text{and} \qquad {}_2D_\phi = -5.8874° $$

and

$$ {}_3D_\theta = 12.9124° \qquad \text{and} \qquad {}_3D_\phi = -5.8875° $$

Further iterations produce no changes greater than 0.0001°, so we take for the blank

$$ D_\theta = 12°55' \qquad \text{and} \qquad D_\phi = -5°53' $$

and since

$$ \theta_{\text{blank}} = \theta_{(hk.l)} + D_\theta \qquad \text{and} \qquad \phi_{\text{blank}} = \phi_{(hk.l)} + D_\phi $$

we have

$$ \theta_{\text{blank}} = 30°46' + 12°55' = 43°01' $$

and

$$ \phi_{\text{blank}} = 19°06' + (-5°33') = 13°33' $$

Thus the value of θ for this hypothetical blank is about 9° too large and the value of ϕ is about 8° too small. Even in this case adequate convergence is obtained in three iterations. If D_θ and D_ϕ are larger, more iterations are required but there is little reason for blanks being cut this far from a set of atomic planes.

ERRORS DUE TO MISALIGNMENT OF THE INCIDENT BEAM

In aligning an x-ray system it is extremely important that the incident beam intersect the goniometer axis. We now investigate the errors resulting from failure to do so. We assume, for simplicity, that the atomic planes in the blank are parallel to the surface of the blank. In a perfectly aligned system the incident beam strikes the crystal at point O (Fig. 11.11), which is the center of the goniometer circle G of radius R. The goniometer reading is θ_B.

Now suppose the goniometer axis and the crystal plate are shifted a distance a in a direction perpendicular to the plate. This moves O to O' and G to G'. The beam now strikes the crystal at point P. The angles of incidence and reflection are still θ_B but the angle measured by the goniometer is now angle g instead of angle θ_B. By considering the geometry of the situation we see that

$$\sin g = \frac{SA}{R} \qquad \text{and} \qquad \tan \theta_B = \frac{SA}{AP}$$

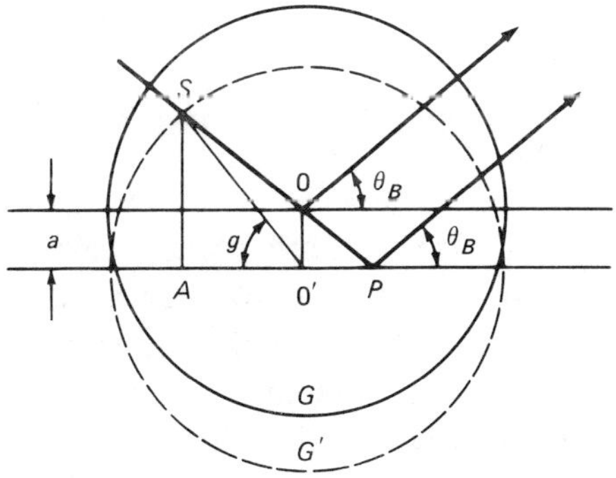

Fig. 11.11. Result of *x*-ray beam not intersecting the goniometer axis.

Therefore

$$\sin g = \frac{AP \tan \theta_B}{R}$$

But

$$AP = AO' + O'P = R \cos g + a \cot \theta_B$$

from which

$$\boxed{\tan \theta_B = \tan g - \frac{a}{R} \cos g}$$

The difference between θ_B and g may be quite significant. For example, suppose the radius of the goniometer circle to be 50 cm and that distance a is 0.1 cm. If the goniometer reading is 15.000°, the true value of θ_B is 14.889°, which is a difference of nearly 7′. Thus a displacement of only 1 mm in the position of the axis of the goniometer, in a typical case, introduces an error of almost 7′ in the angle measurement. This is a common error in making an x-ray diffraction measurement and doubtless accounts for many of the discrepancies between measurements made on different machines.

No error results from failure of the reflecting surface of the crystal plate to contain the goniometer axis O so long as the incident beam passes through O (or would do so if the plate were removed). Thus measurements may be made from either side of a thick plate with parallel surfaces.

DERIVATIONS OF EQUATIONS (107) to (110)

In Fig. 11.8 the incident beam lies in the plane of the goniometer as does the blank normal. The atomic planes make an angle δ with the goniometer axis. If $\delta = 0$, the goniometer reading g is equal to the Bragg angle θ_B and the incident beam, the reflected beam, and the normal to the atomic planes all lie in the goniometer plane.

If $\delta \neq 0$, then the normal to the atomic planes and the reflected beam no longer lie in the goniometer plane although, of course, the incident beam does do so. The plane containing the incident beam, the reflected beam, and the normal to the atomic planes is called the *reflection plane.* The incident beam lies along the intersection of the reflection plane and the goniometer plane. In the reflection plane the angles of incidence and reflection are the Bragg angle θ_B. The goniometer measures angle g, which is the projection of the angle of incidence on the goniometer plane.

We define a coordinate system XYZ in which the goniometer plane is the XY plane and the Z axis is the goniometer axis. (See Fig. 11.12.) The atomic planes intersect the XY plane along the X axis and make an angle δ with the Z axis.

When $\delta = 0$ we define a vector of unit length $\mathbf{V}_1$ along the incident beam and a unit vector $\mathbf{N}_1$ along the normal to the atomic planes, both of course lying in the XY, or goniometer, plane. The components of $\mathbf{V}_1$ are ($\cos\theta_B$, $\sin\theta_B$, 0), and the components of $\mathbf{N}_1$ are (0, 1, 0). Therefore the cross product $\mathbf{V}_1 \times \mathbf{N}_1$ is a vector having components (0, 0, $\cos\theta_B$). The magnitude of this vector, which is $\cos\theta_B$, is the area of the parallelogram formed on $\mathbf{V}_1$ and $\mathbf{N}_1$. (See Fig. 11.14.)

We now assume that $\delta \neq 0$ and define unit vectors $\mathbf{V}_2$ and $\mathbf{N}_2$ both of which lie in the reflection plane. Vector $\mathbf{V}_2$, which lies along the incident beam, has components ($\cos g$, $\sin g$, 0). Vector $\mathbf{N}_2$, which

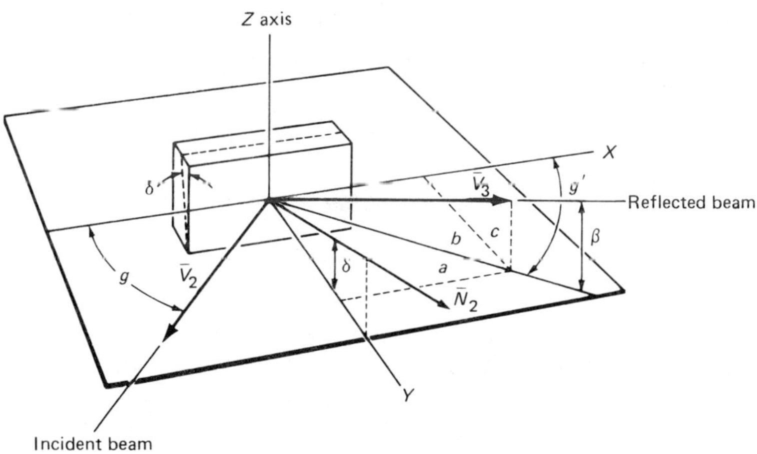

Fig. 11.12. The coordinate axes X and Y lie in the goniometer plane and the axis Z, which contains the goniometer axis, is perpendicular to it. The unit vectors $\mathbf{V}_2$, $\mathbf{N}_2$, and $\mathbf{V}_3$ lie in the reflection plane.

lies along the normal to the atomic planes, has components (0, $\cos\delta$, $\sin\delta$). The vector product[2] $\mathbf{V}_2 \times \mathbf{N}_2$ is a vector having components ($\sin g \sin\delta$, $-\cos g \sin\delta$, $\cos g \cos\delta$). The magnitude of this vector is equal to the area of the parallelogram formed on $\mathbf{V}_2$ and $\mathbf{N}_2$ (in the reflecting plane), which must be equal to the area of the parallelogram formed on $\mathbf{V}_1$ and $\mathbf{N}_1$. This follows from the fact that the angles between both pairs of vectors is θ_B and all of the vectors are unit vectors. Therefore

$$\cos^2\theta_B = \sin^2 g \sin^2\delta + \cos^2 g \sin^2\delta + \cos^2 g \cos^2\delta$$

from which

$$\sin g = \frac{\sin\theta_B}{\cos\delta} \tag{107}$$

We now define a unit vector $\mathbf{V}_3$ lying along the reflected beam and having components (a, b, c). The vector product $\mathbf{N}_2 \times \mathbf{V}_3$ is a vector having components ($c\cos\delta - b\sin\delta$, $a\sin\delta$, $-a\cos\delta$). This vector is the area of the parallelogram formed on $\mathbf{N}_2$ and $\mathbf{V}_3$. It is equal to the vector representing the area of the parallelogram formed on $\mathbf{V}_2$ and $\mathbf{N}_2$. This is true because both areas are equal in magnitude and are coplanar. We thus have the three equations

$$\begin{aligned} c\cos\delta - b\sin\delta &= \sin g \sin\delta \\ a\sin\delta &= -\cos g \sin\delta \\ -a\cos\delta &= \cos g \cos\delta \end{aligned}$$

The last two equations define the value of a, which is

$$a = -\cos g$$

From the first equation $b = c\cot\delta - \sin g$.
We also have, since $\mathbf{V}_3$ is a unit vector,

$$a^2 + b^2 + c^2 = 1$$

---

[2] See page 206 for the definition of the vector (or cross) product.

Eliminating a and b from the last three equations we obtain

$$c = 2 \sin \delta \cos \delta \sin g$$

and using

$$\sin g = \frac{\sin \theta_B}{\cos \delta}$$

we have

$$c = 2 \sin \delta \sin \theta_B$$

The angle g' which determines the detector setting is defined by

$$\tan g' = \frac{b}{a} = \frac{c \cot \delta - \sin g}{- \cos g}$$

which simplifies to

$$\tan g' = \tan g \cos 2\delta \tag{108}$$

The angle β is the angle between the reflected beam and the goniometer plane. Therefore $c = \sin \beta$ and thus

$$\sin \beta = 2 \sin \delta \sin \theta_B = \sin 2\delta \sin g \tag{109}$$

We now consider the effect of rotating the blank by an angle ψ about a normal to the blank. It is assumed that the line of intersection of the atomic planes and the surface of the blank lies in the plane of the goniometer when $\psi = 0$. (In the AT-cut the line of intersection is the crystallographic X-axis.)

We define a unit vector $\mathbf{V}_4$ which lies along the incident beam having components $(\cos g, \sin g, 0)$. Vector $\mathbf{N}_4$ is a unit vector normal to the atomic planes in the blank having components $(\sin \delta \cos \psi,$ $\cos \delta, \sin \delta \sin \psi)$. The vector product $\mathbf{N}_4 \mathbf{X} \mathbf{V}_4$ is a vector having components

$$(\sin g \sin \delta \cos \psi, \cos g \sin \delta \sin \psi, \sin g \sin \delta \sin \psi - \cos g \cos \delta)$$

Just as before, the magnitude of this vector must be equal to $\cos \theta_B$ and therefore

$$\cos^2 \theta_B = \sin^2 \delta \cos^2 \psi + (\cos g \cos \delta - \sin g \sin \delta \sin \psi)^2 \tag{110a}$$

Equation (110*a*) is the exact relationship between the goniometer angle g and the variables θ_B, δ, and ψ. Unfortunately it cannot be solved in closed form for the value of g and must be solved numerically by successive approximations. Fortunately there exists an approximate equation which is easily solved for g and which gives excellent agreement with the exact equation for all reasonable values of δ. The equation is

$$\sin(g + \delta \sin \psi) \doteq \frac{\sin \theta_B}{\cos(\delta \cos \psi)} \qquad (110b)$$

For instance, if we take $\delta = 3.0000°$, $\theta_B = 13.0000°$, and $\psi = 45°$, Eq. (110*b*) gives $g = 10.8878°$ whereas the correct value from Eq. (110*a*) is $g = 10.8868°$ – a difference of about 3″ of arc. The agreement is better for values of ψ near 0°, 90°, 180°, etc., and worse for larger values of δ. The agreement is exact at 90°, 180°, 270°, etc. If we take $\delta = 10°$, $\theta_B = 13°$, and $\psi = 45°$, the value of g from Eq. (110*b*) is 5.994° whereas the correct value from Eq. (110*a*) is 6.0303°, a difference of about 2′ of arc. This is a rather extreme case and it should seldom be necessary to have a difference this large.

DERIVATION OF EQUATIONS (111) AND (112)

Doubly rotated plates present a special case of x-ray orientation and measurement. We now define the equations relating the two goniometer readings with the orientation angles θ and ϕ. Figure 11.13

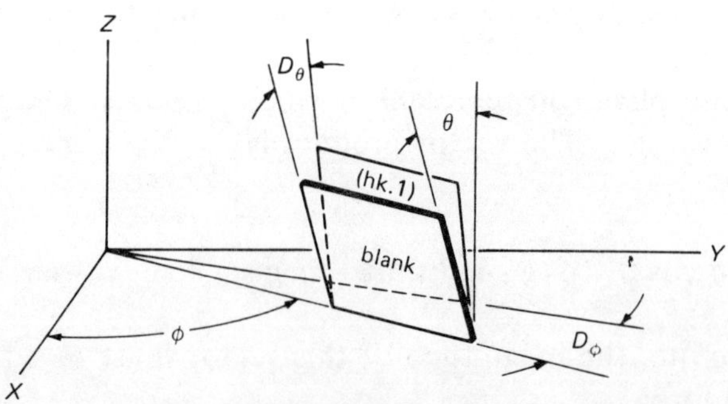

Fig. 11.13. Position of the blank relative to the (*hk*.1) planes. The axes are the crystallographic axes.

shows the positions of the blank and the $(hk.l)$ planes used to orient the blank relative to the crystallographic XYZ-axes.

We let $D_\theta = \theta_{\text{blank}} - \theta_{(hk.l)}$ and $D_\phi = \phi_{\text{blank}} - \phi_{(hk.l)}$

We first calculate the value of g_θ which is the goniometer reading when the blank is in the position for measuring the angle θ (see Fig. 11.12). We assume that the normal to the blank lies in the goniometer plane and is parallel to the Y axis. The $(hk.l)$ planes make angles D_θ and D_ϕ with the surface of the blank, as shown in Fig. 11.13.

We now define a unit vector $\mathbf{N}_5$ perpendicular to the $(hk.l)$ planes. This vector has components (Fig. 11.12)

$$(-\sin D_\theta \cos D_\phi, \cos D_\theta \cos D_\phi, \sin D_\phi)$$

As before, the unit vector $\mathbf{V}_5$ along the incident beam has components

$$(\cos g_\theta, \sin g_\theta, 0)$$

The cross product $(\mathbf{V}_5 \times \mathbf{N}_5)$ is a vector having components

$$(\sin g_\theta \sin D_\phi, -\cos g_\theta \sin D_\phi, \cos g_\theta \cos D_\theta \cos D_\phi + \sin g_\theta \sin D_\theta \cos D_\phi)$$

The magnitude of the vector $(\mathbf{V}_5 \times \mathbf{N}_5)$ is the area of the parallelogram formed on $\mathbf{V}_5$ and $\mathbf{N}_5$ in the reflection plane. This parallelogram has the same area as that formed on $\mathbf{V}_1$ and $\mathbf{N}_1$ in the goniometer plane since the angle between both pairs of vectors is the complement of θ_B. The magnitude of the vector $(\mathbf{V}_5 \times \mathbf{N}_5)$ is $\sin(90^\circ - \theta_B) = \cos\theta_B$ and therefore

$$\cos^2\theta_B = \sin^2 g_\theta \sin^2 D_\phi + \cos^2 g_\theta \sin^2 D_\phi + (\cos g_\theta \cos D_\theta \cos D_\phi + \sin g_\theta \sin D_\theta \cos D_\phi)^2$$

which reduces readily to

$$\sin(g_\theta - D_\theta) = \frac{\sin\theta_B}{\cos D_\phi} \qquad (111)$$

Placing the blank in the position to measure the crystallographic angle ϕ merely interchanges the two angles D_θ and D_ϕ so that by exactly the same reasoning it follows that

$$\sin (g_\phi - D_\phi) = \frac{\sin \theta_B}{\cos D_\theta} \tag{112}$$

In practice the values of g_θ and g_ϕ are measured by x-ray diffraction and the value of θ_B is known. Hence Eqs. (111) and (112) uniquely determine the values of D_θ and D_ϕ the values of which can be calculated with arbitrary accuracy by an iteration method. The orientation angles θ and ϕ for the blank are then found by adding D_θ and D_ϕ to the values of θ and ϕ for the atomic planes.

****

A vector **V** can be expressed in terms of its scalar components a, b, and c as

$$\mathbf{V} = a\mathbf{i} + b\mathbf{j} + c\mathbf{k}$$

where **i**, **j**, and **k** are unit vectors along X, Y, and Z, respectively.

The scalar product of two vectors $\mathbf{V}_1 = a_1\mathbf{i} + b_1\mathbf{j} + c_1\mathbf{k}$ and $\mathbf{V}_2 = a_2\mathbf{i} + b_2\mathbf{j} + c_2\mathbf{k}$ is $\mathbf{V}_1 \cdot \mathbf{V}_2 = a_1a_2 + b_1b_2 + c_1c_2 = \mathrm{V}_1\mathrm{V}_2 \cos\theta$, where V_1 and V_2 are the magnitudes of the vectors and θ is the angle between them.

The vector product of the two vectors $\mathbf{V}_1$ and $\mathbf{V}_2$ is

$$\mathbf{V}_1 \mathbf{X} \mathbf{V}_2 = \begin{vmatrix} \mathbf{i} & \mathbf{j} & \mathbf{k} \\ a_1 & b_1 & c_1 \\ a_2 & b_2 & c_2 \end{vmatrix} = (b_1c_2 - b_2c_1)\mathbf{i} + (a_2c_1 - a_1c_2)\mathbf{j} + (a_1b_2 - a_2b_1)\mathbf{k}$$

$$= \mathrm{V}_1 \mathrm{V}_2 \sin\theta\, \mathbf{n}$$

where **n** is a unit vector perpendicular to $\mathbf{V}_1$ and $\mathbf{V}_2$.

The vector product $\mathbf{V}_1 \mathbf{X} \mathbf{V}_2$ is obviously the area of the parallelogram formed on $\mathbf{V}_1$ and $\mathbf{V}_2$ (see Fig. 11.14).

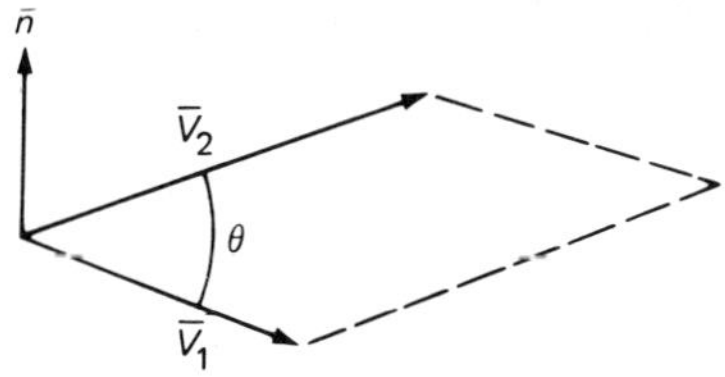

Fig. 11.14. The vector product of two vectors is the area of the parallelogram formed on the two vectors.

12 AGING IN QUARTZ CRYSTAL UNITS

INTRODUCTION

Any change in the characteristics of a device or component with time is called *aging*. Applied to quartz crystal units the term usually means a change in the resonant frequency of the unit, although other parameters, especially the motional resistance, may also change with time. But since the primary function of the crystal unit is to establish and maintain a constant frequency, the term aging is generally taken to imply a change of frequency. Changes in the equivalent motional resistance may influence the amplitude of oscillation in an oscillator circuit or the insertion loss in a filter network but have extremely small effect on the frequency in either case.

The causes of aging in quartz crystal units are many and the effects are complicated and varied. In the following paragraphs we discuss some of the causes and remedies.

AGING DUE TO SURFACE DETERIORATION

The lapped or polished surface of a quartz crystal consists of a layer of quartz which is in a strained condition. Evidence for this is readily obtained by x-ray diffraction. Figure 12.1 is a photographic record of the Bragg reflection from the (10.1) planes of a quartz surface which had been lapped with a 600-mesh SiC abrasive. The first exposure, at the top, was made with the crystal in the correct position for reflection of the Cu $K\alpha$ radiation. For the next exposure the crystal was rotated 2° away from the Bragg position. Successive exposures were made at increments of $\frac{1}{4}$°. It may be seen that reflec-

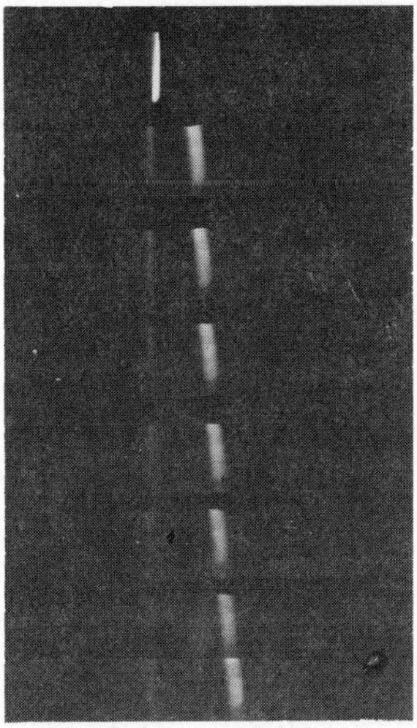

Fig. 12.1. Bragg reflection from a lapped-surface.

tion persists at the Bragg angle even when the crystal is rotated by $3\frac{1}{2}°$ from the Bragg position. This is interpreted to mean that some areas on the surface of the crystal are continually being brought into the Bragg position as the crystal is rotated. The more intense lines at the right in the picture are due to the Bremsstrahlung.

This phenomenon can be understood by considering the lapping operation which is actually a fracturing process. When a grain of abrasive is pressed against the quartz surface, cracks radiate from the point of contact. When cracks completely surround a particle of quartz, the particle falls off. However, at the time that the lapping process is terminated, many partially loosened particles remain. These particles are partially surrounded by boundaries at which the molecular bonds have been broken. Such particles are forced out of alignment with the main body of the crystal and a certain amount of energy is stored in the resulting strain. (The situation is not unlike that of having a piece of food lodged between two teeth, in which case the teeth are pushed out of their proper positions.) As time goes on the cracks continue to grow until they completely surround the particle, which may then fall from the surface. The particles thus dislodged form a dust on the surface of the blank which can be seen with the unaided eye by observing the surface at a glancing angle while dragging a probe across the surface. The size of the dust particles is roughly the same as that of the abrasive used in lapping the surface. The process is greatly accelerated by humidity and high temperatures and is accompanied by drastic increases in the frequency and resistance of the resonator. When the dust is removed by washing, the resistance of the unit returns to normal but the frequency is much higher than before. In extreme cases frequency changes of $0.1f^2$ or more may occur in a period of only a few days.[1]

This type of aging can be reduced somewhat by careful lapping and polishing and by protection from water vapor, but the only complete remedy is etching with HF or NH_4F_2 to relieve the strains causing the misorientation described above. The amount of etching required depends somewhat upon the type of lapping operation but is usually not less than about $0.5\,f^2$ for an AT-cut blank. The amount of etching required for any blank can be determined by making an "etching curve" as shown in Fig. 12.2. The crystal blank is etched under controlled conditions by fixed increments of time and the resulting frequency changes are recorded and plotted against the accumulated etching time. This graph shows that the etching proceeds very rapidly at first but later settles down to a (nearly) linear rate. The rapid initial rate is due to the fact that the chemical process

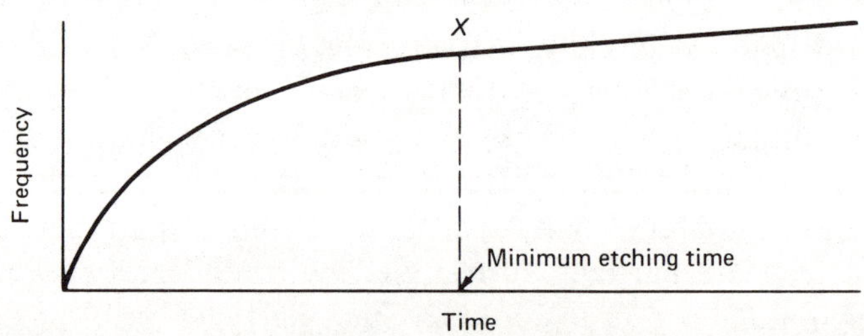

Fig. 12.2. Etching curve for a quartz resonator.

---

[1] It is convenient and customary to express frequency changes in quartz crystal units in units of kilohertz per megahertz². Thus a change of 1500 Hz in the frequency of 10-MHz unit would be called a change of $\Delta f = 1.5$ KHz/100 MHz² $= 0.015\,f^2$. The usefulness of this convention arises from the fact that many effects result in frequency changes which are proportional to the square of the frequency of the unit. It thus provides a normalizing factor.

proceeds most rapidly at points of strain. Etching the quartz until the rate of frequency change with time is linear as indicated by the point *X* in Fig. 12.2 is sufficient to prevent this type of aging. X-ray diffraction photographs made of such etched surfaces show no evidence of strained quartz, but the slightest abrasion is sufficient to cause its return. It is interesting to note that it is not necessary to etch away all the quartz which is initially misaligned but only to relieve the strain by widening the cracks which were produced in the lapping operation. Electron micrographs clearly show the nature of the phenomenon.

AGING DUE TO SURFACE CONTAMINATION

A serious source of aging is contamination of the surface of the blank by volatile materials which may be present in the holder. These materials affect the frequency by loading (and unloading) the surface of the quartz blank. One very serious contaminant is the flux used to make the solder seal in many types of holders. For this reason a great deal of effort has been expended in developing holders which employ other types of seals such as cold welding and capacitance discharge and resistance welding. Each of these sealing methods presents problems of its own. Encapsulation in glass presents a particular problem due to excess heat, especially in miniature-sized units.

Other contaminants which have been identified as causes of aging are residues of volatile materials which are used in attaching the lead wires to the quartz plate and the lubricants used in drawing the metal cans which form the enclosure. Both these contaminants can be minimized by high-temperature baking in vacuum or in a hydrogen atmosphere.

The presence of a contaminating substance generally tends to cause a reduction of the frequency of a resonator, but the effect may be misleading. At any given temperature, equilibrium is eventually established within the enclosure. In this condition each surface is receiving and giving up contaminating material at a constant rate. However, any change in the temperature results in a new equilibrium condition with a different distribution of the contaminants and, in general, a change of frequency. A period of hours or days may be required for the unit to achieve stability at the new temperature. The same thing occurs if the drive level is changed, since this also results in a change in the operating temperature of the quartz piezoid.

The sensitivity of the frequency of a crystal unit to surface contamination can be appreciated by noting that the removal of a single layer of atoms from the surfaces of a 20-MHz AT-cut blank causes a frequency increase of some 100 Hz, or about 5 ppm (5×10^{-6}). Many units are specified to have aging rates of a few parts in 10^8 per year.

AGING ASSOCIATED WITH THE ELECTRODES

Most crystal resonators have thin metal electrodes deposited on the quartz. These electrodes serve two functions: they provide the means of applying the electric field to the quartz and they are used in making the final adjustment of the frequency of the resonator. Thus any changes in the mass of the electrodes resulting from chemical combination with other elements must result in a change in the resonant frequency. The most commonly used metals are silver, gold, and aluminum. Each metal presents a different aging problem.

Silver is used more than any other metal as an electrode material. A thin layer of chromium is often deposited on the quartz beneath the silver to improve adherence. Silver is not attacked by oxygen under ordinary conditions although it is attacked by O_3 or ozone. (The ozone produced by sparks probably contributes to the oxidation of silver contacts in switches.) However, silver is strongly attacked by sulfur and by the halides – fluorine, chlorine, bromine, and iodine. Chemical reaction with any of these and certain other elements results in a decrease in the frequency of the resonator.

Silver iodide is sometimes used in fabricating crystal units. The silver electrodes react instantly at room temperature with iodine vapor, producing a layer of silver iodide on the silver electrodes. The increased mass provides a simple and inexpensive method of adjusting the frequency. The silver iodide thus produced is very stable both physically and chemically and units made by this process can have very good aging characteristics. It is necessary, however, to ensure that no unreacted iodine remains in the enclosure, since subsequent chemical reactions may occur causing further frequency changes and aging. Because complete removal of the excess iodine is difficult, the iodine process is not favored for units with severe aging requirements.

Gold is also extensively used as an electrode material, especially in low-frequency resonators and in those demanding excellent aging characteristics. Gold has the advantage of being chemically inactive and also being easy to deposit. However, it does not adhere well to quartz and its high density and low electric conductivity make it undesirable with high-frequency resonators in which mass loading of the electrodes seriously degrades the performance of the device. Gold has other disadvantages, as we will see later.

Aluminum is commonly used as an electrode material, especially in high-frequency units. Aluminum adheres tenaciously to quartz because of the nature of the silicon to oxygen to aluminum bonding. The acoustic impedance of aluminum closely matches that of quartz, thereby minimizing the effects of acoustic reflection at the quartz-aluminum interface. However, aluminum reacts slowly with oxygen to form Al_2O_3, or sapphire. The result is that aluminum-plated units usually exhibit rather poor aging characteristics with the frequency decreasing roughly logarithmically with time for periods up to a year or more.

The main source of aging in aluminum plated units is the growth of the oxide layer on the electrodes. A newly created aluminum surface begins to oxidize immediately upon exposure to air. The oxide grows until it reaches a thickness of about 50 Å after which the growth substantially ceases. Under ordinary conditions several months is required for the oxide to reach its terminal thickness and the process cannot be accelerated much by increasing the temperature. Therefore, heat treatment of aluminum-plated units is largely ineffective in preventing aging of this type.

It is possible, however, to minimize and to virtually eliminate aging due to oxidation of the aluminum electrodes by the use of anodic oxidation.[2] In this process anodic oxidation is used to create an oxide layer which is initially thicker than the terminal thickness of the natural oxide layer. The thickness of the oxide film is proportional to the voltage used to produce it, thus providing a simple and useful

---

[2] See Method of Adjusting the Frequency of a Crystal Resonator and Reducing Ageing Effects, U.S. Patent 4,130,771, December 19, 1978. This patent discloses the procedure used in adjusting the frequency of an aluminum plated quartz resonator by means of anodic oxidation. See also Virgil E. Bottom, "A Novel Method of Adjusting the Frequency of Aluminum Plated Quartz Crystal Resonators." Proc. *30th Annu. Symp. Freq. Contr.*, June 1976, pp. 249–253.

method for adjusting the frequency while at the same time ensuring freedom from further oxidation.

Aluminum oxide or sapphire is highly inert to most chemical reactions and is physically very durable. The protection thus provided to the aluminum electrodes thereby virtually eliminates aging due to chemical reactions with the electrodes.

A particularly troublesome form of aging is characterized by the refusal of the crystal unit to operate at low drive levels after a period of storage. The phenomenon is known by several names, among which are "second level of drive," "hard starting characteristics," and "sleeping sickness." Typically, after a period of storage, which may be hours, days, or weeks, the crystal unit refuses to oscillate in a circuit that requires the unit to operate at low amplitude of vibration. The unit can often be revived by operating it in an oscillator that drives the unit at high amplitude after which it operates satisfactorily in the low-drive-level oscillator. However, after a period of storage the unit usually repeats the cycle.

The refusal of the unit to operate at a low drive level has been traced, in almost every instance, to failure of the electrode metal to adhere properly to the quartz. Microscopic examination of such units, after having been driven at high amplitude, shows pinholes in the plating where the metal has been thrown off. It is obvious that the electrode metal did not adhere properly to the quartz in these areas. The loose plating, while present, acts as a damping mechanism which reduces the Q of the resonator below the limit required for oscillation. In some cases the metal may have been deposited over a particle of dust, thus forming a blister.

The problem is most severe in crystal units employing silver electrodes applied by evaporation. It also exists with gold-plated units and much less frequently with aluminum-plated units. Unlike aluminum, neither gold nor silver adheres well to quartz and for this reason a thin layer of chromium is often deposited on the quartz before the gold or silver electrodes are deposited. Chromium, like aluminum, forms a SiO_2–oxide–metal bond. The gold or silver then adheres to the chromium instead of to the quartz.

In order to obtain good adherence between the evaporated metal and the quartz surface it is essential that the surface be "clean" and that the metal be deposited under proper conditions. It is difficult to define a clean surface since no surface except one prepared in a perfect vacuum is completely free of adsorbed atoms. But the surface on which the metal is to be deposited must be free of the organic materials which are present in every atmosphere. The subject of surface chemistry, including cleanliness, is too large to be pursued here. It may be said, however, that ultrasonic cleaning with detergents and boiling water followed by ultraviolet light irradiation produces surfaces to which evaporated metals adhere satisfactorily. It is necessary, in addition, to perform the evaporation in a vacuum system which is free of oil and at a pressure no greater than 10^{-6} torr.

AGING ASSOCIATED WITH PHYSICAL EFFECTS

The resonant frequency of a quartz resonator is usually influenced to some degree by any mechanical strain imposed on the piezoid. Consequently, any change in the static stress imposed on the piezoid results in a change in the resonant frequency. Mechanical stresses are imposed upon the AT-cut quartz plate by the mounting structure and by the electrodes, both of which in general have temperature coefficients of expansion different from those of quartz.

Unless great care is exercised in mounting the blank, the supporting leads may exert a tensile force or a bending moment or both on the blank. The frequency of the blank is determined to some degree by the strains resulting from these stresses. It is easy to demonstrate these effects by artificially imposing stresses on an oscillating plate.

All solids, except perfect crystals, undergo stress relaxation with time, especially at elevated temperatures. As the stresses in the mounting structures relax, the frequency of the resonator may increase or decrease depending upon the nature of the stress on the plate. In either case the result is aging.

Stress relaxation in the metal electrodes is a serious source of aging. The frequency of both AT- and BT-cut resonators is strongly affected by radial stresses in the plane of the blank. Thermal stresses associated with the different temperature coefficients of the quartz and electrode metal result in frequency perturbations. After a long time at a fixed temperature, the stresses in the metal relax; but when the temperature is changed, the process is repeated. This is one reason why a crystal unit often starts a new aging cycle when the operating temperature is changed.

It is well known that thin metal films undergo significant molecular rearrangement with time and that these effects are accelerated by high temperatures. Therefore, annealing is useful in minimizing the effects of electrode stresses. The thickness of the film, the metal used and the method of application, the surface to which it is applied, and the presence of intermediate metals all influence the amount of built-in stress, the relief of which causes frequency changes and aging.

The contribution of the electrodes to aging can be eliminated by using a design in which the quartz blank is suspended between conducting surfaces with an air gap between the quartz and the electrodes. However, mechanical problems associated with such designs and their high cost make them unattractive.

AGING ASSOCIATED WITH THE QUARTZ

No quartz crystal is perfect. Impurity atoms, interstitial atoms, lattice vacancies, and nonstoichoimetric ratios are always present in both natural and cultured quartz crystals. Each of these defects has a subtle effect on the elastic coefficients which determine the frequency of the resonator and therefore any change in the distribution of the defects in the crystal may result in aging. The degree of perfection varies from crystal to crystal, both in natural and cultured quartz. For example, the frequency of a BT-cut resonator made from some specimens of natural quartz may be decreased by as much as 0.02 percent by irradiation with x-rays, while a similar resonator made from a different specimen may show little or no change. Those plates which exhibit large frequency changes with radiation also become dark or smoky; those whose frequencies do not change remain uncolored. These phenomena are all understood, but this is not the place to discuss them.

The tendency to change frequency and to darken may be eliminated by "sweeping" (see page 24). Units which must withstand the effects of high radiation fields (such as satellite applications) are usually made from swept quartz. The redistribution of crystalline defects, either from a concentration gradient or an externally applied electric field may induce changes in the elastic coefficients and changes in the frequency or aging.

Relaxation of body strains in crystals resulting from crystalline imperfections can be accelerated by annealing at elevated temperatures, but annealing is a continuous process. Ultimately the distribution of stress effects approaches equilibrium at a given temperature. If the temperature is changed, a new distribution must be established, which is another reason why time is required for a crystal unit to reach frequency stability after a change of temperature.

AGING ASSOCIATED WITH HOLDER DEFECTS

Most crystal holders are back-filled with a dry gas, usually N_2. Often a small percent of He is added to facilitate leak testing with a mass spectrometer. In some units vacuum encapsulating is used and in certain units such as the IT- and SC-cuts, it is essential.

Every holder leaks. It is only a matter of degree. Leaks which permit oxygen, water vapor, and other gases to enter the enclosure may result in changes in the frequency of the device through a number of mechanisms. Exposure to water vapor of an AT- or BT-cut resonator which has been previously vacuum-baked results in a frequency decrease which may be as much as $0.01f^2$. This decrease in frequency is due to the adsorption of a monomolecular layer of water on the previously dehydrated surface. The process is reversible by vacuum-baking the unit again. Some long-term aging effects are doubtless due to this phenomenon.

Unless protected by anodic oxidation, aluminum electrodes may undergo further oxidation with attendant frequency decreases. To avoid further oxidation of aluminum films on which the oxide has not reached the terminal thickness, it would be necessary to remove all oxygen from the holder and prevent the entrance of more. Both of these are virtual impossibilities except with glass holders which have been thoroughly vacuum baked.

Chemical reactions between silver and atmospheric gases such as SO_2 cannot be disregarded as possible sources of aging. Considering the extreme sensitivity of a high-frequency quartz resonator to any mass added to or removed from its surfaces, it is not impossible that the vapor pressure of the various materials in the holder may make some contribution to the aging phenomena. But even if this effect is negligible, no doubt exists that the holder makes a significant contribution to the aging of some crystal units.

ACCELERATED AGING TESTS

As a general rule, aging effects are accelerated by increased temperatures and for this reason it is desirable that the operating temperatures of crystal units be as low as possible. But there is no way in which temperature exactly substitutes for time, nor vice versa. This is obvious if one considers effects such as melting solder or phase changes resulting in broken quartz. A given crystal may remain within specified limits for years at room temperature and fail in a few days at 100°C and instantly at 250°C.

Nevertheless, it is useful and customary to include accelerated aging tests as a part of the specification for a crystal unit. For example, Standard Specification MIL 3098 requires that the frequency of a unit not change by more than 5 ppm in 30 days at 85°C. Some commercial specifications require that the frequency remain within certain limits for a specified time and temperature. As an example, one specification for a 12.0-MHz unit requires that the frequency remain within ±50 Hz of the nominal frequency after 21 days at 120°C.

An accelerated aging test should reflect the use and environmental conditions to be imposed upon the unit. There is questionable value in specifying an accelerated aging test at 100°C for a unit which is to be used at room temperature. It may simply result in the rejection of units which would be quite satisfactory for the purpose and thereby increase the cost.

Aging tests may be made with the units operating or not, but it must not be concluded that the results will be the same. For an aging test to be conclusive, the unit should continuously operate at a fixed temperature and drive level. The frequency may be monitored continuously or checked periodically. Any interruption of the operating conditions will normally result in a discontinuity in the frequency-time curve for one or more of the reasons outlined above. In tests where accelerated aging is interrupted, the units must be maintained at a fixed temperature (say room temperature) for a fixed period of time before recording the frequency. Drive levels must also be standardized. It is unnecessary to emphasize the need for the best test equipment, including ovens and frequency standards in making aging tests. And care must be taken in interpreting the results.

METHODS OF MINIMIZING AGING OF QUARTZ CRYSTAL UNITS

Aging can probably never be eliminated in quartz crystal units, but certain steps can be taken to minimize the effects. The first step is obviously to isolate the cause of the aging. This is complicated by the fact that several sources are usually present simultaneously. It is not uncommon in an aging test for the frequency change to reverse direction as different aging effects predominate and often cancel each other. A high initial rate of aging does not necessarily indicate a long-term effect, nor vice versa.

Substantial decreases in frequency are most often associated with some type of contamination or with a leaking holder. In these cases scrupulous attention to cleanliness and rigorous leak testing are required.

Poor vacuum conditions in plating equipment including inadequate vacuum and contamination from pump oil are common sources. These problems are attacked by ensuring that vacuum pressures are maintained below 10^{-6} torr and by using vacuum systems in which oil is adequately trapped or not used at all.

It is not possible to specify a procedure which will ensure minimum aging in every case. It can be stated that the utmost attention to every detail of the production process is required, since aging may arise from almost any operation in the entire production process.

Nevertheless, a surprising degree of stability is achievable with quartz crystal units. Aging rates as low as 1 part in 10^{11} per day have been achieved. Such a quantity is a bit more comprehensible if it is recalled that the distance across the North Atlantic Ocean is about 5×10^8 cm. A change of 1 part in 10^{11} in this distance is some 0.005 cm. A hypothetical clock controlled by a crystal resonator whose frequency decreases by 1 part in 10^{11} per day would be losing one second per day after running for thirty centuries!

13
THE STRESS–COMPENSATED CRYSTAL UNIT

INTRODUCTION

It has long been known that mechanical stresses applied to a quartz piezoid cause changes in its resonant frequency. A very simple experiment in which a circular AT-cut plate is compressed along a diameter shows that the frequency change is proportional to the applied force and depends on the direction in which the force is applied, even changing sign with the direction. Frequency changes of 100 ppm are easily produced in this way in ordinary quartz resonators.

Another well-known phenomenon is the amplitude-frequency (AF) effect in which the frequency of the resonator varies with the drive level, i.e., the power used in driving it. The frequency shift in a circular AT-cut resonator of conventional design is quite accurately proportional to the square of the driving current and hence to the power dissipated in the quartz. This frequency change is superimposed on, and is much larger than, the frequency change due to the temperature rise of the quartz. It is now recognized that the AF effect is largely, if not entirely, due to thermal stresses in the quartz as the annular ring of cooler quartz constrains the driven portion which is heated by the energy dissipated in it. The fact that the change of frequency occurs almost simultaneously with the change in drive level has led to the hypothesis that the AF effect results from variations of the elastic coefficients with amplitude. Now it can be shown that the resonant frequency of a vibrating system does depend on the amplitude if the stress/strain relationships contain nonlinear terms of odd order. But frequency changes occur in a typical AT-cut plate vibrating at constant amplitude when the central region is heated by exposing it to radiant energy from a hot source or a laser. This proves that the AF effect is not solely due directly to the increased amplitude.

It is instructive to examine the magnitude of the stress which might be expected in a typical case. To do so we may consider a 10 MHz AT-cut plate having electrodes with a diameter of 0.25 cm and a series resistance of 50 Ω. A current of 10 mA through the unit at the series-resonant frequency results in a dissipation of 5×10^{-3} W or 1.2×10^{-3} cal/sec. If this heat were absorbed in the quartz between the two electrodes, its temperature would rise at the rate of about 6°/sec. On page 00 it was shown that a temperature increase of 1°C° in a con-nstrained piece of quartz causes a compressional stress of 12 atms. So we may expect that the temperature rise due to the energy dissipated in the center of the quartz plate will produce very significant stresses. As the central region of the plate tries to expand against the con-straining ring of cooler quartz it tends to produce a bulging effect and for this reason the phenomenon has been called the *oil-can* effect.

A striking example of the oil-can effect occurs in crystal units which are operated in fast warm-up ovens. In some of these ovens the crystal unit is brought to the operating temperature of 60 to 80°C in a period of 1 min or less by means of a heating element placed very near the crystal unit. The thermal stresses resulting from this rapid heating sometimes result in frequency excursions, or "overshoots," of several hundred parts per million during the warmup period. Attempts have been made to heat the quartz quickly by means of heating elements deposited directly on the quartz blank with similar results.

Acceleration forces including that of gravity also influence the frequency of a crystal unit. The frequency changes vary with the method of mounting the piezoid and the direction of the force. The change of frequency with position in the earth's gravitational field is called the *tip-over effect*; in AT-cut plates this is typically a few parts in 10^9 per g.

A closely related phenomenon is the frequency change associated with the stresses applied to the quartz blank by its metal electrodes.

According th EerNisse[1] frequency changes of 3 ppm may result from stresses imposed by the gold electrodes on a 5-MHz fifth-overtone AT-cut resonator. These stresses result from the different coefficients of expansion and the physical changes which occur in metal films with time.

The frequency changes in all of the foregoing phenomena are associated with changes in the elastic coefficients with stress bias. It was shown by the author (see Ref.2) in 1947 that the frequency changes which occur in rotated Y-cut plates upon irradiation to saturation with x-rays are also due to changes in the effective values of the elastic coefficients. These changes are due to internal stresses resulting from irradiation, and in those crystals showing the maximum effect they are as follows:

Maximum change in $c_{14} \sim -0.07\%$
Maximum change in $c_{44} \sim -0.03\%$
Maximum change in $c_{66} \sim -0.02\%$

These values were deduced from the frequency changes which are observed in AT-, BT- and Y-cut plates the values of which are, respectively, < 0.001 percent, -0.02 percent, and -0.01 percent. The observation that the various stress effects are strongly anisotropic suggests that some orientation might exist for which the frequency-stress effect might be zero, at least for certain kinds of stresses.

HISTORY OF THE SC-CUT

On the basis of theoretical calculations, EerNisse (Ref. 1) predicted that a doubly rotated quartz plate cut at the angles of $\phi = 22.5°$ and $\theta = 34.3°$ (see Fig. 11.10) should be free of frequency changes due to stress bias in the plane of the plate. In 1976 EerNisse (Ref. 3) extended the calculations to include frequency changes due to any mechanical stress bias. One important result of EerNisse's calculations was the conclusion that the frequency-stress coefficient should depend almost entirely upon the angle ϕ, while the first-order frequency-temperature

[1] See Refs. 1 and 3 at the end of this chapter.

coefficient should be determined almost entirely by the angle θ. He called the new piezoid the *SC-cut* (stress-compensated).

In 1977, Kusters et al. (Ref. 4) showed experimentally that the results predicted by EerNisse were obtained with a plate having the orientation angles $\phi = 21.93°$ and $\theta = 34.11°$. They called it the TTC-cut (thermal-transient-compensated) and pointed out its advantages as well as some of the problems associated with its production and use.

The nomenclature TS-cut [used by Holland (Refs. 12 and 13) in 1964], TTC-cut [used by Kusters (Ref. 4)], and SC-cut all refer to a plate cut at angles near $\theta = 34°$ and $\phi = 22°$; the exact angles depending on the application. We use the term SC-cut (stress-compensated), which is the one most favored at this time.

The orientation of the SC-cut is similar to that of the IT-cut introduced in 1951 by Ives and Bottom (Ref. 5). The IT-cut was developed to obtain a unit having a zero frequency-temperature coefficient at a temperature higher than that of the AT-cut. The orientation of the IT-cut is $\phi = 19°06'$ and $\theta = 34°17'$. The value of ϕ is the same as that of the (12.2) planes, thereby making x-ray orientation of the IT-cut no more difficult than that of the AT- and BT-cuts. It was not noted at the time that the frequency-stress effect is much less in the IT- than in the AT-cut, although its relative freedom from activity dips and spurious modes was pointed out.

Further theoretical treatments of, and experimental results on, the SC-cut are given in papers by Ballato et al. (Ref. 6), and Gagnepain et al., (Ref. 7).

FABRICATION OF THE SC-CUT

The problem of cutting an accurately oriented quartz plate such as an *X*-, *Y*-, or *Z*-cut in which the plane of the plate is normal to one of the crystallographic axes is relatively simple. It is somewhat more complicated if the plane of the plate is rotated about one of the axes, as with the AT- or BT-cut. It is more difficult if the plane of the plate is rotated about two of the crystallographic axes as in the IT-cut and becomes still more difficult if the line of intersection of the plate and a set of atomic planes does not lie in one of the crystallographic planes. This is the case with the SC-cut.

The *X*-, *Y*-, and *Z*-axes are readily identified in quartz. Cultured quartz is supplied with accurately oriented *X*- and *Z*-surfaces ground on lumbered *Y*-bars. In the following it is assumed that we use a *Y*-bar prepared in this way.

We begin by defining a rectangular coordinate system whose axes coincide with the *X*-, *Y*-, and *Z*-crystallographic axes, as shown in Fig. 11.10. We assume the saw to be oriented in such a way that the saw plane is normal to the *Y*-axis.

We now define a unit vector **N** which is normal to the plate having coordinates ϕ and θ (Fig. 13.1). The components of this vector are

$$\begin{pmatrix} -\cos\theta\sin\phi \\ +\cos\theta\cos\phi \\ \sin\theta \end{pmatrix}$$

The problem then is to find the angles through which the bar (or the saw) must be rotated to bring the unit vector **N** into coincidence with the *Y* axis of the coordinate system. A method for doing so has been described by Bond and Kusters (Ref. 8) and the following procedure is based upon this method.

We have shown in Chap. 2 that the matrix of Eq. (10*b*) performs a rotation of a vector about the *Y* axis and that the matrix of Eq. (10*c*)

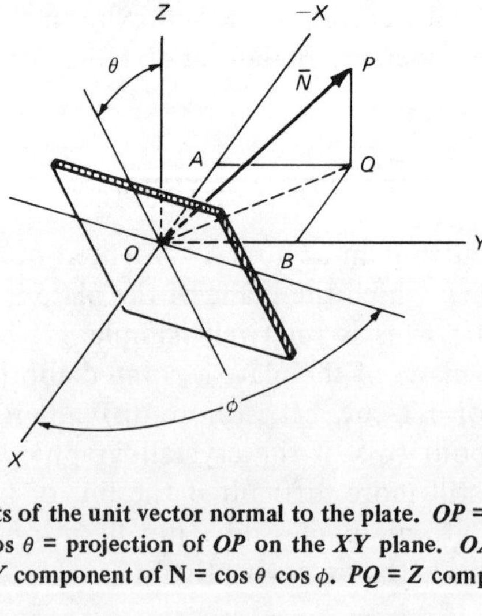

Fig. 13.1. Components of the unit vector normal to the plate. *OP* = N = unit vector normal to the plate. *OQ* = cos θ = projection of *OP* on the *XY* plane. *OA* = *X* component of N = - cos θ sin φ. *OB* = *Y* component of N = cos θ cos φ. *PQ* = *Z* component of N = sin θ.

performs a rotation about the Z axis. We will use these two matrices to perform the two successive rotations required to bring the vector **N** into coincidence with the Y axis. We let β be the angle by which the bar must be rotated about the Y axis and γ the angle by which it must be rotated about the Z axis. The operation for performing the two rotations is the matrix product

$$\mathbf{N}'' = \begin{pmatrix} \cos\gamma & \sin\gamma & 0 \\ -\sin\gamma & \cos\gamma & 0 \\ 0 & 0 & 1 \end{pmatrix} \begin{pmatrix} \cos\beta, & 0 & -\sin\beta \\ 0 & 1 & 0 \\ \sin\beta & 0 & \cos\beta \end{pmatrix} \begin{pmatrix} -\cos\theta\ \sin\phi \\ \cos\theta\ \cos\phi \\ \sin\theta \end{pmatrix}$$

Carrying out the indicated operations gives the components of **N**″, which are

$$N''_X = -\cos\theta\ \sin\phi\ \cos\beta\ \cos\gamma - \sin\theta\ \sin\beta\ \cos\gamma + \cos\theta\ \cos\phi\ \sin\gamma$$
$$N''_Y = +\cos\theta\ \sin\phi\ \cos\beta\ \sin\gamma + \sin\theta\ \sin\beta\ \sin\gamma + \cos\theta\ \cos\phi\ \cos\gamma$$
$$N''_Z = -\cos\theta\ \sin\phi\ \sin\beta + \sin\theta\ \cos\beta$$

Now if **N**″ is to be a unit vector parallel to the Y axis, then $N''_X = N''_Z = 0$ and $N''_Y = 1$. Setting $N''_Z = 0$ gives

$$\tan\beta = \frac{\tan\theta}{\sin\phi}$$

which defines the required rotation β about the Y axis.

Multiplying the equation for N''_X by $\sin\gamma$ and the equation for N''_Y by $\cos\gamma$ and adding the two (not forgetting that $N''_Y = 1$), we easily get

$$\cos\gamma = \cos\theta\ \cos\phi$$

which defines the required rotation about the Z axis. These two equations tell us how to rotate a quartz bar from the initial position in which the coordinate and crystallographic axes coincide to bring it into the position required to saw a blank having the orientation angles ϕ and θ.

The process is quite simple. The lumbered Y-bar is first rotated about the Y axis by the angle β and held in this position by some means, while the assembly is rotated about the Z axis by the angle γ. A cut made in the XZ plane of the original coordinate system has the desired orientation.

Suppose that it is desired to cut an SC-wafer having angular coordinates

$$\theta = 34°07' = 34.117° \quad \text{and} \quad \phi = 21°56' = 21.933°$$

From the equations above,

$$\beta = 61°08' = 61.13° \quad \text{and} \quad \gamma = 39°49' = 39.83°$$

We use a lumbered Y-bar of right-hand quartz having accurately oriented surfaces normal to the X- and Z-axes. The long dimension of the bar is along the Y-axis and the intersections of the lumbered surfaces are parallel to the crystallographic Y-axis.

The bar is placed on a glass or ceramic plate with the crystallographic axes parallel to the coordinate axes and with the $+X$-axis pointing toward the operator. The top surface, which is an XY-surface, is painted for later identification. The bar is then rotated *away* from the operator by an angle of 61°08′ and held in this position by a suitable wax or plaster. The bar and the plate to which it is attached are then rotated *clockwise*, as seen from above, by the angle 39°49′ and then secured to the transfer plate or the saw table. Wafers are then sawed from the bar by making cuts perpendicular to the original position of the Y-axis. With reasonable care and currently available equipment, wafers can be cut within ± 15′ or closer to the desired orientation. If necessary the wafers can be corrected later to closer tolerances.

The wafers cut from the Y-bar have the shape of parallelograms. Two sides of the parallelogram lie in the XY-crystallographic plane and the other two in the YZ-plane. The edge of the wafer which lies in the XY-plane is identified by the previously applied paint and designated the *reference edge*. Next, two directions are identified on the face of the wafer, i.e., parallel and perpendicular to the reference edge.

The wafer is placed on the x-ray chuck with the reference edge parallel to the goniometer axis and D_θ is measured. The wafer is then rotated 90°, making the reference edge perpendicular to the goniometer axis and D_ϕ is measured. From these two measurements the values of θ and ϕ are calculated using Eqs. (111), (112), and (113). Care must be exercised to make the measurements on the correct side of the blank and with the reference edge in the correct position. Other measurements must be treated in a slightly different manner. If the x-ray equipment is set up to use the (12.2) planes (and this is highly desirable), the values of D_θ and D_ϕ will each be about 3°.

If it is necessary to measure the orientation of the circular SC-cut blanks, the direction of the reference edge must be maintained. This may be done by grinding a flat edge, accurately parallel to the reference edge, on the blank.

A more complex method for determining the orientation of doubly rotated blanks has been described by J. Clastre et al. (Ref. 11). The reader is referred to that paper for the details.

It is useful in visualizing the SC-cut wafer to make a model of a lumbered *Y*-bar from a piece of wood of rectangular cross section. A slab to simulate the SC-cut can be cut from the model by giving it the proper rotations and sawing a slab with a miter saw.

The necessity for making two rotations in orienting the bar for sawing SC-cut wafers may be avoided by using SC-bars especially grown for the purpose. A description of the SC-bar has been given by Ballato (Ref. 9). Such bars are quite expensive, and the advantage of requiring only a single rotation does not appear to justify the additional cost. Checking the orientation of wafers cut from the SC-bar is at least as difficult as checking those cut from *Y*-bars. The number of wafers and blanks obtained per kilogram of *Y*-bar quartz is approximately the same for both SC- and AT-cuts.

CHARACTERISTICS OF THE SC-CUT

The SC-cut was developed to obtain a resonator whose frequency would be independent of stresses in the plane of the vibrating plate resulting from thermal stresses and electrode changes. However, it has proved to have a number of other advantages over the conventional AT- and BT-cut plates. Among these are:

1. Superior warmup characteristics due to reduction of the effects of stresses associated with thermal gradients in the plate.
2. Reduced AF effect for the same reason.

3. Almost complete freedom from activity dips.
4. A flatter frequency-temperature curve, especially at the turning points.
5. Reduced tipover effect.
6. Better short-term stability due to reduced response to thermal transients.
7. Reduced aging associated with relaxation of electrode stresses.
8. Relative immunity to the effects of radiation.
9. Less aging due to changes in mounting stresses.

The added complexities of orientation and cutting are disadvantages of the SC-cut but these have not proven to be as difficult as originally perceived. Another disadvantage is that the SC-cut is very sensitive to air damping and must therefore be operated in vacuum to obtain optimum Q values.

Some differences between the SC-cut and the AT-cut may be advantageous or not depending upon the application. The inflection point of the frequency-temperature curve is about 70° higher in the SC- than in the AT-cut. The temperatures of the turning points can be varied by changing the angle θ just as in the AT-cut (see Figs. 8.3 and 8.4). However, changes are required simultaneously in the angle ϕ in order to maintain the stress compensation feature.

The SC- like the IT-cut has three thickness modes of vibration which are designated the *a, b,* and *c* modes. (These piezoids also have two low-frequency face-shear modes.) The *a* mode is a quasi-longitudinal mode analogous to the thickness mode of an *X*-cut plate. It has a frequency constant of about 3370 kHz mm and a temperature coefficient of about -52° ppm/°C. The *b* mode is a quasi-shear mode having a frequency constant of about 1990 kHz mm and a T_f of about -25 ppm/°C. Within the range 0 to 100°C the frequency-temperature curve of the *b* mode is almost linear. The *c* mode is also a quasi-shear mode. It is analogous to the AT cut. The frequency constant of the *c* mode is about 1800 kH mm and consequently the SC-cut plate is about 10 percent thicker than an AT-cut plate of the same frequency.

The relative activities of the *b* and *c* modes can be varied a bit by the design of the plate but usually they are about the same. The SC-cut plate can be excited on either mode or on both modes simultaneously. The presence of the two modes requires special attention in the design of oscillator and filter circuits employing the SC-cut plate. Both modes are accompanied by the usual family of inharmonic overtone modes.

The reason for the presence of the three thickness modes in the SC- and IT-cut resonators can be understood by considering the piezoelectric strains produced by an electric field in a quartz plate. For simplicity we consider three plates, one cut perpendicular to the X-axis, one perpendicular to the Y-axis and one at an intermediate position. The projections of these three plates on the Z-plane are shown in Fig. 13.2. In all cases the electric field is perpendicular to the plate.

In the X-cut (a) the electric field has components along the X_2 and X_3 axes as well as along X_1. The effects of these components is to produce an extensional (positive) strain along X_1 and compressional (negative) strains along the direction perpendicular to X_1. These are responsible for the piezoelectric coefficients d_{11} and $d_{12} = -d_{11}$ [see Eq. (26)].

The electric field in the Y-cut plate (b) has a component along X_1 producing a positive strain and one along X_3 causing a negative strain. There is no component along X_2. The resultant of the two longitudinal strains is a shear strain (see page 41) and the piezoelectric coefficient $d_{26} = -2d_{11}$.

The rotated blank (c) in which the applied electric field has components along all three of the X-axes experiences both longitudinal and shear strains resulting in the a and c modes. The shear strain which results in the b mode is orthogonal to that which produces the c

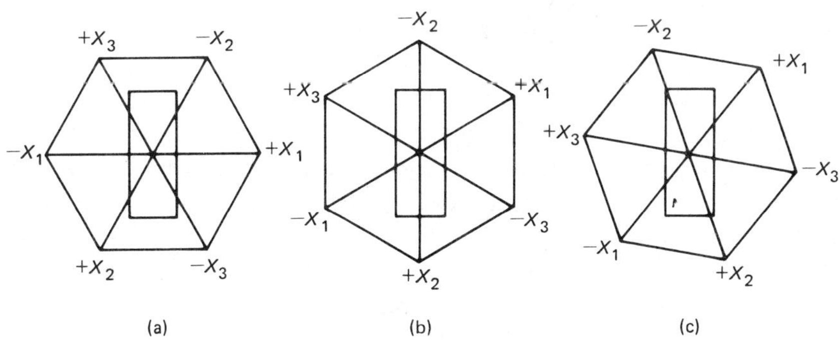

Fig. 13.2. Piezoelectric strains in X-cut, Y-cut, and rotated plates.

mode. It occurs as a result of the θ rotation and therefore would not be present in the simple model of Fig. 13.2*c*.

The reason for the sensitivity of the SC-cut to air damping can also be understood by comparing the strains produced by the electric field in the *Y*-cut plate of *b* and those induced in the rotated plate of *c*. In the *Y*-cut plate the positive strain parallel to X_1 can be resolved into components along X_2 and perpendicular to X_2. The same can be done for the negative strain along X_3. The resultant of these strains is zero in the direction perpendicular to the plate. In the rotated cut, however, the components of the strains along X_1, X_2, and X_3 do not cancel in the direction perpendicular to the plate, with the result that the surface suffers a displacement in this direction. The resulting pumping action is responsible for the air damping observed in the SC-cut.

TESTING SC-CUT UNITS

The chief difficulty in testing SC-cut crystal units is the presence of the *b* mode. A CI-meter, tuned to the frequency of the *c* mode, may have a tendency to jump to the *b* mode. Thus the CI-meter is not well adapted to making frequency-temperature tests on the SC-cut. The transmission-line method using a phase-locked loop is somewhat more reliable but, of course, more expensive and less convenient. Oscillators having narrow-band filters in the feedback network are indicated for use in test equipment for SC-cuts.

The frequency of the SC-cut is less sensitive to the rate of temperature change and therefore frequency-temperature tests can be carried out more rapidly than with units containing AT- or BT-cut plates.

Since the shape of the frequency-temperature curve in the SC-cut is largely determined by the value of the angle θ, a frequency-temperature test is adequate for determining the correct value for this angle. But the frequency-stress behavior is largely a function of the angle ϕ, and to determine the correct value for this angle, some type of frequency-stress test is required.

It is necessary to distinguish between thermal-stress sensitivity and *g* or acceleration sensitivity. Different tests are required. A simple test for thermal-stress sensitivity is to abruptly change the temperature of the crystal unit and to observe the time required for the unit to achieve frequency stability. The same result can be achieved by suddenly changing the drive level. A method (Ref. 12) has been described for testing the g sensitivity which appears to lend itself to production as well as laboratory testing. This method will now be described.

The crystal unit is rigidly attached to a vibration table calibrated for both frequency and acceleration. The frequency of the crystal unit is modulated by the acceleration-induced stresses through the frequency-stress effect. As in any FM signal, sidebands appear in the output of the crystal-controlled oscillator the frequencies of which are given by

$$f_n = f_0 \pm n f_m \qquad n = 0, 1, 2, 3, \ldots$$

where f_0 is the unmodulated (carrier) frequency and f_m is the modulating frequency i.e., the frequency of the vibration table. The amplitudes of the sidebands are given by $A_n = A\, J_n\, (\Delta f/f_m)$, where $\Delta f = S f_0 X$. X is the acceleration and S is the fractional change in frequency per g. J_n is Bessel's function of order n.

If $(Sf_0X/f_m) = F \ll 1$, then $J_0\,(F) \doteq 1$ and $J_1(F) \doteq F/2$. Therefore

$$\frac{A_1}{A_0} = \frac{(Sf_0X)}{2f_m} \quad \text{and} \quad S = \left(\frac{2f_m}{Xf_0}\right)\left(\frac{A_1}{A_0}\right)$$

The ratio A_1/A_0 can conveniently be determined by the use of a spectrum analyzer with a logarithmic scale and since f_0, f_m, and X are known, the value of S is readily determined.

As an example, suppose that a 10-MHz unit is vibrated at a frequency of 500 Hz with a maximum acceleration of 100 g and that the ratio of the amplitude of the first side band to that of the carrier is 0.001. Then

$$S = \frac{2\ (500)\ 0.001}{10^7\ 100} = 10^{-9} \text{ per g}$$

The value of $F = [10^{-9}\ (100)\ 100^7]/500 = 0.002$, which is certainly small compared with unity, thus justifying the approximation used in the theory.

REFERENCES

1. E. P. EerNisse, *Proc. 29th Annu. Symp. Freq. Contr.*, 1975, pp.1-4. (Includes 21 references.)
2. V. E. Bottom, *Phys. Rev.* **71**:476 (1947). (Abstract.)
3. E. P. Eernisse, *Proc. 30th Annu. Symp. Freq. Contr.*, 1976, pp. 8-10. (Includes 16 references.)
4. J. Kusters, C. Adams, H. Yoshida, and J. Leach, *Proc. 31st Annu. Symp. Freq. Contr.*, 1977, pp. 3-7.
5. V. E. Bottom and W. R. Ives, U. S. Patent No. 2,743,144, April 24, 1951.
6. A. Ballato, E. P. EerNisse, and T. Lukaszek, *Proc. 31st Annu. Symp. Freq. Contr.*, 1977, pp. 8-16. (Includes 24 references.)
7. J. J. Gagnepain, J. C. Ponçot, and C. Pegeot, *Proc. 31st Annu. Symp. Freq. Contr.*, 1977, pp. 17-20. (Includes 8 references.)
8. Walter L. Bond and John A. Kusters, *Proc. 31st Annu. Symp. Freq. Contr.*, 1977, pp. 153-158. (Includes 11 references.)
9. A. Ballato, *Proc. 33d. Annu. Symp. Freq. Contr.*, pp. 322-336. (Includes 105 references.)
10. A. Warner, B. Goldfrank, M. Meirs, and M. Rosenfeld, *Proc. 33d Annu. Symp. Freq. Contr.*, 1979, pp. 306-310.
11. J. Clastre, C. Pegeot, and P. Y. Leroy, *Proc. 32nd Annu. Symp. Freq. Contr.*, 1978, pp. 310-319.
12. R. Holland, *1974 Ultrasonic Symp. Proc.*, IEEE Catalog 74 CHO 896-1 SU, pp. 592-598.
13. R. Holland, *IEEE Trans. Sonics and Ultrasonics,* **SU-21**:171-178 (1974).

APPENDIX 1
BASIC DIELECTRIC THEORY

Consider a parallel-plate capacitor of area A with a quartz plate of thickness e as a dielectric. When the terminals of the capacitor are connected to a battery, the charging current which flows is composed of three parts:

1. The Maxwell displacement current
2. The dielectric polarization current
3. The piezoelectric polarization current

If the dielectric were vacuum instead of quartz, only 1 would flow. If the dielectric were glass, 1 and 2 would be present. But if the dielectric is a piezoelectric material such as quartz, then all three are present.

The first current

$$i_1 - A \frac{dD_1}{dt}$$

is called the Maxwell displacement current. D_1 is called the Maxwell displacement. It is the charge per unit area which appears on the plates of the capacitor. The displacement is related to the electric field between the plates by the equation

$$D_1 = \epsilon_0 E \qquad (a)$$

where D_1 and E have the same direction and ϵ_0 is called the permittivity of vacuum.

The second current is the dielectric polarization current. We may think of a dielectric as being composed of + and - charges with limited freedom of motion relative to each other. Each molecule (or other unit) is said to have an induced dipole moment p when the field is applied. The dipole moment of each mole-

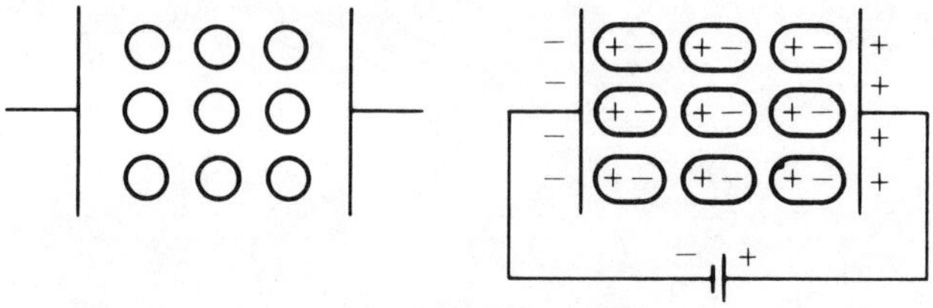

Dielectric without field   Dielectric with field

cule is the product of the magnitude of the charge and the distance between the two charges. If we have n dipoles per unit volume, the dipole moment per unit volume, or the polarization, is

$$P = n\,p$$

If we let Q be the charge per unit area induced on the electrodes by the polarization of the dielectric, then QA is the charge on one electrode. The other electrode has charge $-QA$. The dipole moment of the dielectric is therefore QAe. But Ae is the volume of the dielectric, so that Q is the dipole moment per unit volume or the polarization. Hence we have the important result that the polarization of a dielectric is equal to the induced surface charge per unit area. This is true whether the polarization is due to dielectric or piezoelectic effects or both.

When the capacitor is charged the dielectric is said to be polarized. It should be noted that the polarization P which is taken to point from the negative charge toward the positive charge of the dipole is opposite to the direction of the electric field which points from the positive toward the negative electrode.

In many dielectric materials the polarization and hence the electric displacement is proportional to the applied field, so we may write

$$-P_2 = D_2 = \eta E \qquad (b)$$

where η is called the dielectric susceptibility.

The third current is the piezoelectric displacement current resulting from the polarization due to the strain in the crystal. It is found that the polarization in quartz is proportional to the strain, so that we may write

$$P_3 = \epsilon\, x$$

But the strain is related to the electric field by the equation

$$x = d\,E$$

Therefore the polarization and electric displacement are given by

$$-P_3 = D_3 = d\epsilon E \tag{c}$$

Combining Eqs. (*a*), (*b*), and (*c*) we have

$$D = D_1 + D_2 + D_3 = (\epsilon_0 + \eta + d\,\epsilon)\,E = k'E$$

or

$$D = \epsilon_0 \left(1 + \frac{\eta}{\epsilon_0} + \frac{d\epsilon}{\epsilon_0}\right) E$$

The quantity $1 + \eta/\epsilon_0$ is called the dielectric constant in the case of an ordinary dielectric, i.e., one which is not piezoelectric. It is usually designated K_e. The quantity $1 + \eta/\epsilon_0 + d\,\epsilon/\epsilon_0$ is analogous to the dielectric constant in an ordinary dielectric.

The force acting on a charge q in an electric field is

$$F = -qE$$

where the minus sign signifies that the force on a positive charge is opposite in direction to E.

The work done in moving a charge q a distance dy parallel to E is

$$F\,dy = -q\,E\,dy$$

The work per unit charge, which is $-E\,dy$, is called the potential difference dV. Therefore $dV = -E\,dy$ and

$$E = -\frac{dV}{dy} \tag{d}$$

Let us now suppose that the charge density in the dielectric is not constant but varies from point to point. This is the case if the dielectric is also piezoelectric and the strain is not uniform. We suppose the strain and therefore the charge density, q_v, to vary in the y direction. Let $q_v\,(y)$ represent the charge density expressed as a function of y. Consider an element in the form of a thin slice Δy. Let E be the field intensity at the left side of the element. Then the field intensity at the right side is

$$E + \frac{dE}{dy}\,\Delta y$$

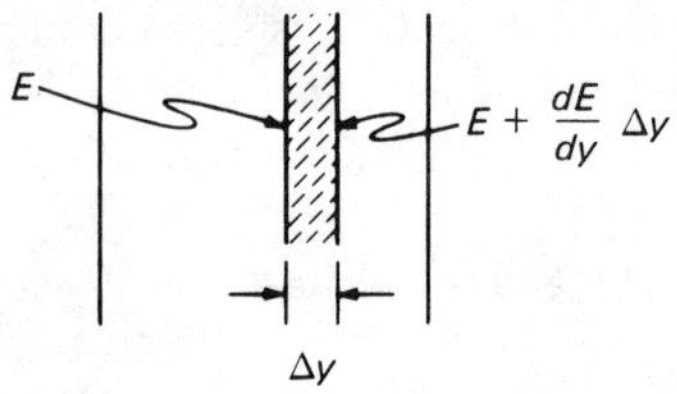

By Gauss' law

$$k\,\epsilon_0 \int_s \mathbf{E} \cdot d\mathbf{A} = \int_v q_v(y) A\, dy$$

On the left side of the element the surface integral yields $k\epsilon_0 A E$ directed toward the left. On the right side it yields

$$k\,\epsilon_0 A \left(E + \frac{dE}{dy}\,\Delta y\right)$$

directed toward the right. The integrals over the other sides of the element are zero, since the field is assumed parallel to the y direction. Therefore the integral over the entire surface of the element is

$$k\,\epsilon_0 A \frac{dE}{dy}\,\Delta y$$

But the charge enclosed by the element is simply $q_v A \,\Delta y$ and therefore by Gauss' law

$$k\,\epsilon_0 \frac{dE}{dy} = q_v(y) \tag{e}$$

We may note, in passing, that if $q_v = 0$, then $dE/dy = 0$ and therefore E is constant. But if q_v is not zero, then E is a function of y. This is the case when the strain varies in the y direction due to vibration in a piezoelectric material.

The electric displacement $D = k\,\epsilon_0\, E$, and therefore

$$\frac{dD}{dy} = k\epsilon_0 \frac{dE}{dy}$$

From (*e*),

$$\frac{dE}{dy} = \frac{q_v}{k\epsilon_0} \tag{f}$$

Therefore,

$$\frac{dD}{dy} = q_v(y)$$

Since P is oppositely directed to D and E

$$\frac{dP}{dy} = -q_v(y) \qquad (g)$$

From (d),

$$\frac{dE}{dy} = -\frac{d^2V}{dy^2}$$

By using (f) we have Poisson's equation:

$$\frac{d^2V}{dy^2} = -\frac{q_v}{k\epsilon_0} \qquad (h)$$

If $q_v = 0$ we have LaPlace's equation

$$\frac{d^2V}{dy^2} = 0 \qquad (i)$$

Equations (g), (h), and (i) are required in deriving the equivalent circuit of the quartz resonator in Chap. 6.

APPENDIX 2
COMPLEX NUMBERS

The algebra of complex numbers is used to handle problems involving steady-state alternating currents and voltages. More advanced techniques are required to treat transient conditions, but we are not concerned with these in this book.

An imaginary number is the product of a real number and the square root of -1, which we designate by j. For example, $4j$ is an imaginary number. A complex number is the vector sum of a real number and an imaginary number. For example, $z = 2 + 3j$ is a complex number. Complex numbers may be represented graphically by marking off the real part of the number on the horizontal, or real, axis and the imaginary part on the vertical, or imaginary, axis. The magnitude of the complex number z is indicated by $|z|$ and if $z = a + bj$, then $|z| = (a^2 + b^2)^{1/2}$.

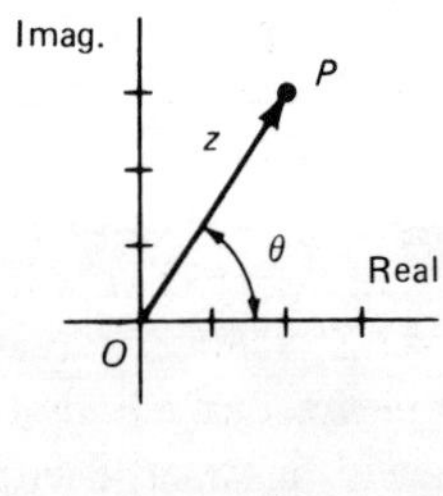

The vector OP used to represent z makes an angle $\theta = \tan^{-1}(b/a)$ with the real axis.

The sum of two complex numbers is found by adding the real parts and adding the imaginary parts, i.e., if $z_1 = 2 + 3j$ and $z_2 = 4 - 5j$, then $z_1 + z_2 = 6 - 2j$.

The product of two complex numbers is found by multiplying the two numbers as binomials and writing $j^2 = -1$. Thus $(2 + 3j)(4 - 5j) = 8 + 12j - 10j - 15j^2 = 23 + 2j$.

Division by complex numbers is not defined. A fraction having a complex denominator is rationalized by multiplying the numerator and denominator by the complex conjugate of the denominator and then simplifying. For example,

$$\frac{2+3j}{3+4j} = \frac{2+3j}{3+4j}\,\frac{3-4j}{3-4j} = \frac{18+1j}{3^2+4^2} = \frac{18}{25} + \frac{1}{25}\,j$$

A complex number multiplied by j is another complex number and the vectors representing the two are perpendicular to each other. For example, $j(3 - 2j) = (2 + 3j)$. The effect of multiplication by j is to rotate the vector counterclockwise by 90°. Multiplying twice by j rotates the vector by 180°. This is equivalent to multiplication by -1.

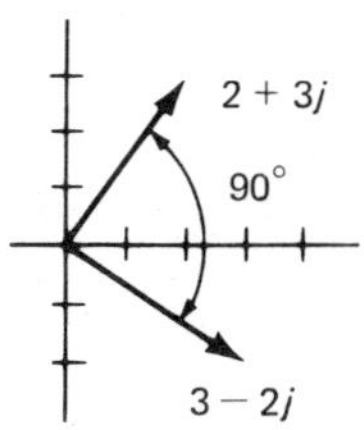

Complex numbers may also be represented in polar form by giving the length (or modulus) of the vector and the angle made with the real axis. Thus $z = 3 + 4j$ may be represented by a vector of length 5, making an angle θ $\tan^{-1}$ (4/3) with the real axis, and written $z = (r, \theta)$, with $r = 5$ and $\theta = 53^\circ 8'$.

Euler's formula, which can be written

$$z = \exp(j\theta) = \cos\theta + j\sin\theta$$

associates the exponential and trigonometric forms of a complex number. The exponential form is widely used in problems involving alternating currents.

By using complex numbers to represent currents, voltages, and impedances, Kirchhoff's laws can be applied to ac circuits as well as dc circuits. Thus instead of writing $V = I\,R$, we write $V = I\,Z$, where it is understood that all three of the quantities are complex. The impedance of a circuit consisting of a resistance R, an inductance L, and a capacitance C, in series, is written as

$$Z = R + j\omega L - j\,\frac{1}{\omega C}$$

where $\omega = 2\pi f$ is called the angular frequency and is measured in radians per second.

Now if an ac voltage $e = E_m \sin(\omega t)$ is impressed on the circuit, the current which flows is given by

$$i = \frac{e}{Z} = E_m \frac{\sin(\omega t)}{R + jX}$$

where X is written for $\omega L - 1/\omega C$.

Rationalizing the denominator as described above we have

$$i = \frac{E_m (R - jX)}{R^2 + X^2} \sin(\omega t)$$

or

$$i = E_m \left(\frac{R}{R^2 + X^2} - j \frac{X}{R^2 + X^2} \right) \sin(\omega t)$$

Now the amplitude of the current is itself a complex number, which means that the vector representing the current is not parallel to the vector representing the voltage. This is equivalent to saying that the current and voltage are not in phase. If we think of the vectors as rotating counterclockwise, and if $(\omega L - 1/\omega C)$ is positive, then the current i lags behind the voltage e by an angle $\theta = \tan^{-1}(X/R)$. If $(\omega L - 1/\omega C)$ is negative, the current i leads the voltage by the same angle; and if $(\omega L - 1/\omega C) = 0$, the current and voltage are in phase with each other.

It is often more convenient mathematically to represent a current or voltage by its exponential form rather than the trigonometric form and to write

$$i = I_m \exp(j\omega t)$$

instead of

$$i = I_m \sin(\omega t) \quad \text{or} \quad I_m \cos(\omega t)$$

We may think of the quantity $\exp(j\omega t)$ as a vector of unit length which makes one complete rotation about the origin in time $2\pi/\omega$. The projection of this vector is equivalent to the projection of a point moving with motion $\sin(\omega t)$.

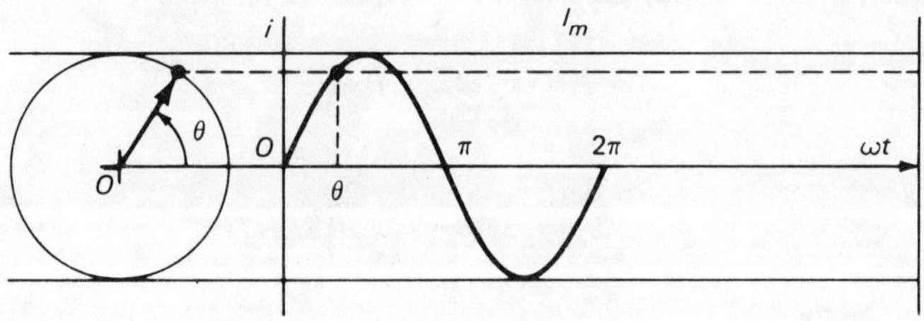

APPENDIX 3
MEASURING TECHNIQUES

INTRODUCTION

The CI-meter provides the most convenient method for determining the resonant frequency and the equivalent resistance of the main mode of the quartz resonator, but it is not suitable for determining the frequency or equivalent resistance of the inharmonic and other weakly excited modes of vibration.

Information concerning these modes, which is often required of units designed to operate as elements in filter circuits, may be obtained by one of the following methods.

BRIDGE METHODS

A circuit which is especially useful in detecting and measuring the resistance of weak modes is shown in Fig. *a*.

T is a ferrite core transformer balanced with respect to ground and broadly tuned to the crystal frequency by C_1. The signal generator supplying the input to the bridge must be capable of being adjusted within a few hertz of the frequency

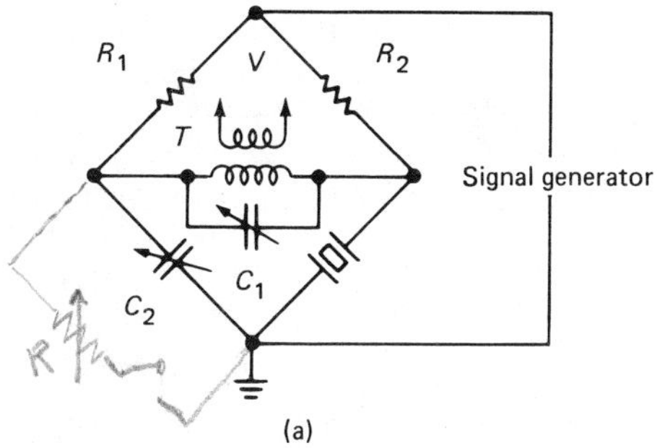

(a)

of the unit under test. It must have excellent frequency stability and it is helpful if the output voltage is variable over a wide range. The Hewlett-Packard Signal Generator Model 606A is easily modified to meet all these requirements by adding a vernier to the fine-frequency control. A still better source is one of the excellent frequency synthesizers which are now available.

R_1 and R_2 are each 100 Ω, roughly matching the generator impedance. The bridge is balanced at a nonresonant frequency of the crystal unit by making C_2 equal to C_0, the static capacitance of the crystal unit. The change in impedance at the resonant frequency disturbs the balance of the bridge, as indicated by the vacuum-tube voltmeter V.

The bridge circuit is especially useful in searching for weakly excited modes, since the admittance of the static capacitance C_0 is balanced out by means of C_2. The resistance in parallel with C_2, which restores the balance of the bridge at the resonant frequency, is equal to the series-resonant frequency resistance R of the crystal unit.

PI-NETWORK METHODS

The Pi-network of Fig. *b* has certain advantages for determining the frequency and equivalent resistance of easily excited modes. V_1 and V_2 are vacuum-tube voltmeters of suitable range. The Hewlett-Packard Signal Generator Model 606A provides the input voltage. R_1 and R_2 are each 50 Ω. It is essential to use a voltmeter at the input terminals, since the voltmeter in the 606A reads correctly only when the terminal impedance is 50 Ω.

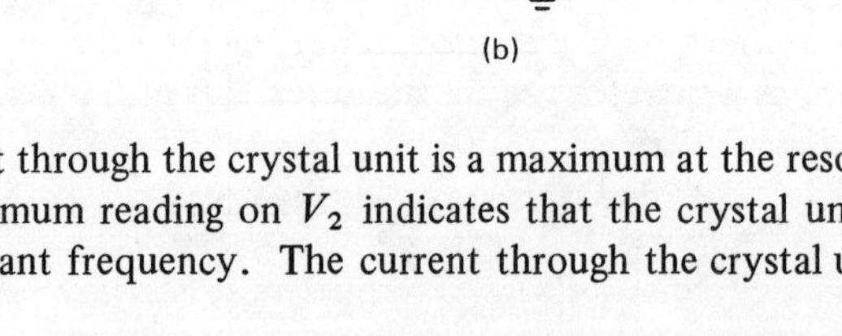

(b)

The current through the crystal unit is a maximum at the resonant frequency. Hence, a maximum reading on V_2 indicates that the crystal unit is operating at its series-resonant frequency. The current through the crystal unit is then given by

$$i = \frac{V_1}{R + R_2}$$

provided that the equivalent resistance $R \ll 1/\omega C_0$. In this case,

$$V_2 = iR_2 = V_1 \frac{R_2}{R + R_2}$$

from which

$$R = R_2 \left(\frac{V_1}{V_2} - 1 \right)$$

The Pi-network has the advantage that the equivalent resistance is easily obtained provided $R \ll 1/\omega C_0$. For weak modes, R is likely to be of the order of $1/\omega C_0$ or even greater, in which case the method is not applicable. The modes may not even be observable, because the current through the series arm of the equivalent circuit is small compared with the current through the static capacitance C_0.

It is possible to identify weak modes using the Pi-network by placing a variable inductance across the crystal unit and using it to tune C_0 to resonance. It is simpler, however, to use the bridge circuit of Fig. *a*.

VECTOR VOLTMETER METHODS

The vector voltmeter provides a very straightforward (but expensive) method of determining all the parameters of the crystal unit and it can be used on weak and strong modes. The circuit used is shown in Fig. 7.4. The series-resonant frequency is the frequency at which the vector voltmeter indicates zero-phase angle. The voltmeter may be used to measure V_1 and V_2 and R can be calculated using the equations above.

The antiresonant frequency of the crystal unit with a load capacitance C_x may be determined by placing C_x in series with the crystal unit and using the vector voltmeter to determine the frequency of zero phase. This frequency is the same as the parallel-resonant frequency. See page 108 ff.

An important improvement is the addition of a phase-locked loop to control the generator frequency, thus eliminating the tedious task of adjusting the frequency of the generator to the resonant frequency of the crystal unit. In this way the vector voltmeter (or transmission line) method may be used for making temperature runs as well as for routine measurements of frequency.

ZERO-PHASE RESISTORS

Adjustment and calibration of vector voltmeters is done by substituting a resistor for the crystal unit and tuning the circuit to zero phase. It is assumed that the phase angle of the resistor is zero. At frequencies above a few megahertz, carbon resistors are usually inductive. For example, a 0.1 W carbon resistor, at a frequency of 10 MHz, may have a positive phase angle of 10°.

Noninductive resistors may be prepared by putting small, high-quality capacitors in parallel with the resistors. The parallel combination should be mounted in a crystal holder identical to that used with the crystal unit which is to be tested. The impedance of the parallel combination can be measured with a suitable bridge or RX meter at the frequency of the crystal unit. The socket used in the measurement should be identical to that used in the vector voltmeter. By selection and adjustment, a combination of carbon resistor and capacitor can be found which yields the desired resistance at zero phase at the nominal frequency of the crystal unit. Measurements should be made with the can in place, and after the correct values are found, the can should be sealed and marked with the resistance and frequency.

It was shown on page 116 [Eq. (79)] that the frequency of a typical 10 MHz unit changes about 1 Hz/degree. Hence, the measurement of frequency made against a 10° resistor may be in error by 10 Hz, or about 1 ppm.

APPENDIX 4
THE PIEZOELECTRIC COUPLING COEFFICIENT

An electromechanical transducer is a device which converts electric energy to elastic (or mechanical) energy, and vice versa. The quartz crystal does so through the piezoelectric effect. When a voltage V is applied to the electrodes of a large quartz plate of area A and thickness e ($A \gg e^2$), the energy stored electrically is given by

$$W_e = \frac{1}{2} C_0 V^2 = \frac{1}{2} k \epsilon_0 \left(\frac{V^2 A}{e}\right)$$

k is the dielectric constant of the unclamped crystal and therefore W_e is the sum of the dielectric and piezoelectric strain energies.

The piezoelectric strain energy per unit volume in the crystal is

$$\frac{1}{2} c x^2 = \frac{1}{2} \left(\frac{\epsilon^2}{c}\right) \left(\frac{V}{e}\right)^2$$

The piezoelectric strain energy in the crystal is therefore

$$W_p = \frac{1}{2} \left(\frac{\epsilon^2}{c}\right) \left(\frac{V^2 A}{e}\right)$$

where ϵ is the applicable piezoelecric constant and c is the applicable stiffness coefficient.

By definition the coupling coefficient κ is given by $\kappa = \sqrt{W_p/W_e}$, so

$$\kappa = \sqrt{\frac{\epsilon^2}{c k \epsilon_0}}$$

The coupling coefficient κ is closely related to the capacitance ratio r. From page 98 we have

$$r = \frac{C_0}{C} = \frac{\pi^2 k c \epsilon_0}{8 \epsilon^2}$$

and therefore

$$r = \frac{\pi^2}{8}\left(\frac{1}{\kappa^2}\right)$$

Note: In the derivation above the difference between the dielectric constants of the free and unclamped plates has been neglected for simplicity. When the difference is included

$$r = \frac{\pi^2}{8}\left(\frac{1}{\kappa^2} - 1\right)$$

The difference is negligible in quartz although not necessarily so with other materials.

The value of κ can be computed for the AT-cut by using the applicable values. The applicable piezoelectric constant is $\epsilon'_{26} = \epsilon_{14} \sin\theta \cos\theta - \epsilon_{11} \cos^2\theta$. With the values of ϵ_{11} and ϵ_{14} from page 62, and $\theta = 35^\circ$

$$\epsilon'_{26} = 0.097\ C/\text{m}^2$$

The value of the applicable stiffness coefficient $c'_{66} = 29.3 \times 10^9$ N/m^2 (page 142). The applicable dielectric constand is $k'_{33} = 4.54$ (page 35). Using these values the piezoelectric coupling coefficient for the AT-cut plate is found to be

$$\kappa = 0.089 \qquad \text{and} \qquad \kappa^2 = 0.0080$$

The capacitance ratio for the AT-cut plate is therefore

$$r = \frac{C_0}{C} = \frac{\pi^2}{8\kappa^2} = 154$$

This is the theoretical value for an infinite plate where it is assumed that the strain is uniform over the entire surface of the plate; that no fringing of the field occurs at the electrode edges; and that no stray holder capacitances exist. None of these conditions is satisfied in an actual resonator, so that typical values measured on a well-designed AT-cut resonator operating on its fundamental mode are $r = 200$ and $\kappa^2 = 0.006$.

The value of r is proportional to the square of the harmonic order. Thus the value of r for an AT-cut resonator operating on its third-harmonic overtone mode is typically 2000 or more. At the fifth harmonic r is greater than 5000.

The antiresonant frequency of a crystal unit is given by

$$f_A = f_R \left(1 + \frac{C}{2C_t}\right)$$ see page 123.

From this equation it is obvious that f_A can be varied by changing C_t with a load capacitance. The degree to which f_A can be varied by changing C_t is called *pullability*. The pullability obviously depends on the value of the motional capacitance C. A crystal unit having a small C is said to be a "stiff" crystal unit.

The frequency of a crystal unit is often "trimmed" to a specified value by varying C_x (and therefore C_t). In temperature-controlled oscillators (TCXOs) C_x is varied by means of a semiconductor diode whose capacitance can be changed by means of a biasing voltage.

For maximum frequency stability a crystal unit should have a small motional capacitance, i.e., be a "stiff" crystal. For maximum pullability the motional capacitance should be large.

It is readily seen that κ^2 should be small for the greatest frequency stability. Physically this means that if the piezoelectric coupling coefficient is small, changes in the parameters of the electric circuit have minimum effect upon the mechanical properties of the piezoid.

APPENDIX 5
GLOSSARY

Every field of human endeavor has its own specialized vocabulary. Common words are often used in a sense which is unfamiliar to those outside the field of specialization. Much of the difficulty of learning a new subject is due to the unfamiliar vocabulary. This is especially true for those not familiar with the language who must rely upon a dictionary which is usually of little help.

The definitions below explain the meaning of the words or terms as they are used in the field of applied piezoelectricity.

activity – A qualitative measure of the amplitude of the vibration of the quartz resonator when operated in a particular oscillator circuit.

activity dip – A decrease in the activity associated with a change of temperature of a crystal unit.

amplitude-frequency effect (also AF effect) – The change in the frequency of a quartz crystal unit with the amplitude of the drive level.

AT-cut – A plate cut from a crystal of quartz in such a way that the plate contains the *X*-axis and makes an angle of about $35\frac{1}{4}°$ with the optic axis on the *r* side.

aging – The change of frequency or other characteristics of a quartz crystal unit with the passage of time.

alpha quartz – The crystalline form of SiO_2 at temperatures below 573°C.

antiresonance – The frequency at which a parallel combination of reactive elements presents a reactance of zero. The impedance at the antiresonant frequency is usually quite large.

axis – A direction (not a line) in a crystal. The axes are chosen arbitrarily in such a way as to make the description of the physical properties of the crystal as simple as possible.

b mode – The mode of vibration in a doubly rotated crystal plate such as the IT- or SC-cut in which the primary displacement is normal to the face of the plate.

beta quartz – The crystalline form of SiO_2 at temperatures above 573°C.

blank – A plate of quartz usually cut at certain specified angles with respect to the crystalline axes and intended for use in fabricating a quartz resonator.

BT-cut – A plate cut from the quartz crystal in such a way that the plate contains the *X*-axis and makes an angle of about 49° with respect to the optic axis on the *R* side.

c mode – (see *b* mode) The mode of vibration in a doubly rotated crystal plate such as the IT- or SC-cut in which the primary displacement is parallel to the surface of the plate. The frequency of the *c* mode is lower than that of the *b* mode.

contour – (v.) The process of generating spherical surfaces on quartz plates. (n.) The surface curvature; usually expressed in dipoters.

crystal – A solid body composed of atoms, molecules, or ions arranged in a regular three-dimensional array. (The term is often used interchangeably with *quartz crystal unit*.)

crystal unit – An electric component consisting of a quartz piezoid, electrodes, and a protective cover.

cultured quartz – Crystalline SiO_2 produced in a factory as opposed to natural quartz provided by nature.

Curie temperature – The temperature at which quartz changes from the alpha to the beta phase. The temperature is 573°C at atmospheric pressure.

cutoff frequency – The lowest frequency which can be transmitted by a wave in a given wave guide or acoustic structure.

dimensioning – The process of fabricating a piezoid to a precisely determined set of dimensions for the purpose of avoiding activity dips. Also predimensioning.

diopter – A term used in the optical industry to measure the magnifying power of a lens. The laps used to generate spherical surfaces on lenses are marked in diopters. The diopter rating of a lens is the reciprocal of its focal length which, for a planoconvex lens, is given by $D = 1/f = (n - 1)(1/R)$. If R is the radius of curvature and $n = 1.5$, which is typical for glass, then $D = 1/2R$, where R is measured in meters.

energy trapping – A term applied to the application of the cutoff phenomenon in wave guides to suppress undesired modes of vibration in quartz crystal units.

face – One of the natural surfaces which develop on a crystal during the growth process. Also *natural face*.

face shear – A mode of strain in which a square plate is distorted into the shape of a rhombus.

frequency – The number of mechanical vibrations executed in 1 sec measured in Hertz (Hz) = 1 cycle/sec; kilohertz (kHz) = 10^3 Hz; megahertz (MHz) = 10^6 Hz.

IT-cut – A doubly rotated quartz plate in which $\theta \doteq 34°20'$ and $\phi = 19°6'$. See Fig. 11.10.

lumbered (bar) – A crystal of (cultured) quartz on which have been cut surfaces which are accurately oriented with respect to the crystallographic axes.

piezoid – A body of some special shape cut from a crystal having piezoelectric properties and used as an electromechanical transducer.

plate-back – (v.) The process of reducing the frequency of a piezoid to the desired or nominal frequency by depositing electrodes on its surfaces. plate-back (n.) The amount by which the frequency is reduced.

quartz – Crystalline SiO_2. Usually refers to alpha quartz.

r face – One of the three smaller faces which occur at the ends of the natural quartz crystal.

R face – One of the three larger faces which occur at the ends of the quartz crystal.

reference edge – The edge of a blank or wafer identified for use in orienting the blank or wafer on the x-ray chuck for making orientation measurements.

reference flat – A flat edge on an otherwise circular blank for use as a reference edge. Called an *X*-flat when perpendicular to the *X*-axis.

resonator – A piece of quartz cut in such a way that it can be excited into mechanical vibration at a certain frequency.

resonator frequency – (1) The frequency of one of the normal modes of a vibrating system; an eigenfrequency. (2) The frequency at which a series combination of reactive elements has a zero reactance. The impedance of a series-resonant circuit is usually low.

rhomb face – One of the *r* or *R* faces on the crystal of quartz.

SC-cut – A doubly rotated quartz plate in which $\theta \doteq 34°7'$ and $\phi \doteq 21°56'$ See Fig. 11.10. Also TTC-cut and TS-cut.

seed crystal – The plate or bar on which the crystal of cultured quartz is grown.

spurious mode – One of the undesired modes of vibration of a piezoid. Usually refers to one of the inharmonic overtone modes.

spur – Contraction of *spurious mode*.

swept quartz – Quartz from which certain impurities have been removed by the use of an electric field.

test set – A standardized oscillator used for measuring the parameters of a crystal unit.

twinning – The reversal of the direction of some physical property of a crystal. Optical twinning in quartz occurs when the plane of polarization of light is rotated in opposite directions in parts of the same crystal. Electrical twinning occurs when the direction of the polar axis is reversed in different parts of the same crystal.

wafer – A slab of quartz sawed from the crystal but not yet shaped to form a blank or a resonator.

X-cut – A surface sawed perpendicular to the *X*-axis. Also a plate or wafer cut perpendicular to the *X*-axis.

Y-bar – A cultured crystal of quartz grown in such a way that its long dimension is parallel to the *Y*-axis. Used for cutting AT-, BT-, and SC-cut wafers.

Y-cut – See *X*-cut.

Z-cut – See *X*-cut.

ZZ' angle – The angle between the plane of the blank and the optic axis in quartz.

APPENDIX 6
PROBLEMS AND QUESTIONS

CHAPTER 1

1. Calculate the change of frequency caused by the removal of one layer of atoms (5×10^{-8} cm) from both sides of a 10 MHz AT-cut blank whose thickness is 0.0169 cm.
2. List the properties of (alpha) quartz which makes it so valuable in the field of electronic communications.
3. The crystal unit in a watch is designed to have a frequency of 2^{15} Hz. If it actually has a frequency of $2^{15} + 1$, how much will the watch gain per day?

CHAPTER 2

1. Explain why the quartz should not be subjected to thermal or mechanical stresses during the fabrication of piezoelectric resonators.
2. The ratios of the intercepts of a given plane with the *X*-, *Y*-, and *Z*-axes of a cubic crystal are 2:1:3, respectively. What are the Miller indices of the plane?
3. Show that the algebraic sum of the first three digits in the B-M indices of a plane in quartz is always zero.
4. Compute the reciprocal of the following matrix

$$A = \begin{pmatrix} 1 & 3 & 4 \\ 2 & 0 & 7 \\ 1 & 5 & 1 \end{pmatrix}$$

5. The following gives the thermal conductivity of quartz in cal/cm sec deg.

	$Z \parallel$	$Z \perp$	
-50°	40	20.5	
0°	32	17.0	$\times 10^{-3}$
+50°	25.5	14.9	

Compute the thermal conductivity in the direction parallel and perpendicular to a line making an angle of 35° with the *Z*-axis at a temperature of 25°C.

CHAPTER 3

1. A plate of quartz is cut with the following specifications: 2.00 cm × 2.00 cm × 0.0500 cm in the *X*-, *Y*-, and *Z*-directions, respectively. The blank is laid on a flat level surface and a 1-kg brass weight is placed on top of it. Assume that the weight is evenly distributed and compute (*a*) the applied stress, (*b*) the resulting strain, and (*c*) the change of thickness.
2. The coefficients of linear expansion for quartz are 14.3×10^{-6} and 7.8×10^{-6} per degree in the directions perpendicular and parallel to *Z*, respectively. Compute the dimension change in the quartz plate of Prob. 1 when the temperature is changed from 0 to 100°C.
3. What stress must be applied to a *Z*-cut plate of quartz to maintain the strain at zero when the temperature of the plate is changed by 1°?
4. A quartz rod having a length of 10 cm and a diameter of 0.5 cm is cut from a quartz crystal. The length of the rod is in the *X*-direction. A tensile force of 10 N (about 2.2 lb) is applied to the rod. Compute the dimensional changes which result. (Be sure to use the correct *c* and *s* constants.)

CHAPTER 4

1. A circular *X*-cut plate has a radius of 1.00 cm and a thickness of 1.00 mm. The major surfaces are plated. The plate is subjected to a compressional force equal to the weight of 1 kg (9.8N). Assume that the force is evenly distributed over the surface and that the electric insulation is perfect. Compute the voltage developed across the blank. Procedure:

    1. Compute the stress.
    2. Compute the strain.
    3. Compute the polarization.
    4. Compute the capacitance.
    5. Compute the voltage.

2. A potential of 1000 V is applied to the electrodes of the quartz plate of Prob. 1. What compressional force must be simultaneously applied to avoid any change in the thickness of the plate? Procedure:

    1. Compute the electric-field intensity.
    2. Compute the piezoelectric strain (no constraint).
    3. Compute the stress required to reverse the strain.
    4. Compute the required force.

3. Verify the two equations at the top of page 62.

4. A potential difference of 1 V is applied to the electrodes of an X-cut quartz plate having a thickness of 0.0100 cm. How much does the thickness of the blank change? What temperature change would be required to produce the same change of thickness?

CHAPTER 5

1. The velocity with which a wave is propagated on a string is given by $v = (T/m)^{1/2}$, where T is the tension and m is the mass per unit length of the string. Compute the frequencies of the fundamental and overtones of a violin string assuming reasonable values for the tension T, the length L, and m.
2. The wave velocity in a square drumhead is 50 m/sec. The drumhead has dimensions of 75 × 75 cm. Compute the frequencies of the fundamental and the first nine overtones. Explain why the theoretical and experimental values differ.
3. A metal rod having a length of 75 cm and a diameter of 0.5 cm is rigidly clamped at its midpoint. The velocity of a longitudinal wave in the rod is 1500 m/sec. Compute the frequencies of the tones produced by striking the bar on the end.
4. Explain why the "frequency constant" (frequency × thickness) of an actual AT-cut plate is not constant.
5. Figure 5.6 is the experimentally observed overtone spectrum of a contoured 3.1-MHz AT-cut plate having a diameter of 1.4 cm. Use Eq. (42) and calculate the overtone frequencies for an isotropic plate of the same size. The effective radius of the contoured plate is about one-half that of the plate. Arrange the calculated frequencies in series as in Fig. 5.6 and note the differences due to neglect of the effects of anisotropy.

CHAPTER 6

1. Obtain Eq. (46) by substituting Eq. (45) into Eq. (44).
2. A circular X-cut blank having a diameter of 1.50 cm and a thickness of 0.3000 mm is fully plated. Find the frequency and the approximate values of C_0, L_m, and C_m. Discuss the errors resulting from the various approximations and assumptions.
3. Repeat the calculations of Prob. 2 with a Y-cut plate having the same dimensions.
4. Repeat the calculations of Prob. 2 with an AT-cut and a BT-cut having the same dimensions.

CHAPTER 7

1. A certain crystal unit has a series-resonant frequency f_R of 10.000000 MHz and a series resistance of 20 Ω. The static capacitance $C_0 = 2.50$ pF. When operated with a load capacitance of 7.5 pF the frequency $f_A = 10.002500$ MHz. Determine the values of L, C, and Q for this unit. ($L = L_m$, $C = C_m$.)
2. What will be the antiresonant frequency of the crystal unit of Prob. 1 when operated with a load capacitance of 32 pF?
3. A resistor having a phase angle of 10° is used in setting a vector voltmeter to zero phase. How much error is introduced into the measurement of the frequency of a 10-MHz unit having a Q of 1×10^5 by using this resistor?
4. Two crystal units have the same series-resonant frequency $f_R = 10.000\,000$ MHz. The motional capacitances are 0.00675 pF and 0.00850 pF, respectively. Each unit is operated in oscillator A in which $C_x = 7.5$ pF and in oscillator B in which $C_x = 10.0$ pF. The static capacitances of the two units are equal to 2.5 pF. Find the frequencies at which the two crystals operate in each oscillator. (This is the principle source of oscillator correlation problems.)
5. Determine the six critical frequencies of the crystal unit of Prob. 1 for (*a*) a load capacitance of 7.5 pF and (*b*) for a load capacitance of 22.5 pF.

CHAPTER 8

1. A quartz blank having a thickness of $0.46\overline{8}$ mm oscillates at a frequency of 5450 kHz. What is its probable orientation? How could one verify this conclusion?
2. A quartz blank having a thickness of $0.006\overline{2}$ in oscillates at a frequency of 31.55250 MHz. What is the probable orientation? How could one verify the conclusion?
3. Discuss qualitatively the factors which determine the frequency-temperature relationship in a crystal unit. Explain how the T_f is minimized in the AT-cut.
4. Refer to Figs. 8.3 and 8.4 and note that the curves for AT-cut blanks cut at 35°22′ are quite different. Explain the difference.
5. A 5-MHz AT-cut unit is to be used within the temperature range 0 to 60°C. What is the best orientation for this unit? What angular tolerances are compatible with a frequency tolerance of ± 5 ppm from the room-temperature frequency?
6. If the correct value for the angle θ is substituted in Eq. (92), the value of T_f can be made equal to zero. Yet the curves of Fig. 8.4 show that even if T_f is zero at room temperature, it is not zero at other temperatures. Explain.

CHAPTER 9

1. Explain the origin of activity dips (band breaks) in thickness-shear quartz resonators. What can be done to avoid them?
2. Why are activity dips less troublesome in units designed to operate in the frequency range 20 to 50 MHz than in units designed to operate in the 2 to 5 MHz range?
3. Refer to the *c* matrix for quartz (13b, p. 45) and use it to explain why AC- and BC-cut plates exhibit fewer activity dips than do AT- and BT-cut plates.

CHAPTER 10

1. Explain the origin of the inharmonic overtone modes in quartz resonators.
2. Explain why the frequency separation of the inharmonic modes increases as the size of the vibrating area is reduced. (No mathematics is required.)
3. Design a crystal unit to operate at 20 MHz on its fundamental mode over the temperature range -40 to +90°C having no spurs within 100 kHz of the main mode and the largest possible motional capacitance. Use an AT-cut blank.
4. What are the effects of contouring the surface(s) of an AT-cut quartz plate? What are the reasons for this behavior?
5. What is the radius of curvature of a surface generated by a lap marked 1.25 diopters?

CHAPTER 11

1. Calculate the lattice spacing and the Bragg angle for the (12.2) planes.
2. Find the angular separation between the Cu $K\alpha_1$ and Cu $K\alpha_2$ radiations reflected from the (12.2) planes of quartz.
3. Explain why the x-ray beam reflected from an etched quartz surface is weaker than that reflected from the same lapped surface before etching.
4. If the (20.3) planes were used to check the orientation of an AT-cut plate, what goniometer readings would be observed? *Ans.*: 48°32′, 19°34′, and 35°20′.
5. Explain quantitatively why one or both of the beams reflected from the plate of Prob. 4 with the *X*-axis horizontal may not be detected.
6. The (12.2) planes are used to check the wafer sawed to make an SC-cut blank. The observed goniometer readings are g_θ = 37°19′ and g_ϕ = 36°26′. What is the orientation of the wafer? *Ans.*: θ = 34°10′ and ϕ = 21°35′.

CHAPTER 12

1. The specifications for a certain 7500-kHz quartz plate require that it be etched by at least 0.5 f^2. How much must the frequency be increased by etching?

2. Show that the thickness of the plated electrode which results in a plate back of $n f^2$ is independent of the frequency f.
3. If a precision crystal unit has a frequency of 5.000 000 000 MHz and an aging rate of -2×10^{-10} per day what will its frequency be at the end of one year? How many cycles will it have lost during the year?
4. Suppose that a layer of water 2.5 Å thick is adsorbed on each side of a 10-MHz AT-cut resonator. How much will this change the frequency? *Ans.*: About 10 Hz.
5. The density of aluminum is approximately equal to that of quartz. If a film thickness of 1500 Å is required for adequate electric conductance, show that the minimum permissible plate back with aluminum on an AT-cut resonator is about $0.2f^2$.

CHAPTER 13

1. Calculate the angles γ and β by which a lumbered Y-bar must be rotated to saw a wafer for the IT-cut orientation.

BIBLIOGRAPHY

This bibliography is not intended to be exhaustive. The intent is to provide references to a few of the most important papers and books which should assist the beginner to find his way into the subject of piezoelectricity. Those references which include extensive bibliographies are indicated by an asterisk*.

Bond, Walter L. "The Mathematics of the Physical Properties of Crystals," *B.S.T.J.*, **22**:1–72 (1943).

*Bottom, Virgil E. "Piezoelectric Effect and Applications in Electrical Communication," *Proc. I. R. E.*, **50**(5):929–931 (1962).

Buchanan, John P. "Handbook of Piezoelectric Crystals for Radio Equipment Designers," ASTIA Document AD 110448, 1956.

*Cady, W. G. "The Piezoelectric Resonator," *Proc. I. R. E.*, **10**:83–114 (1922).

*Cady, W. G. "Bibliography on Piezoelectricity," *Proc. I. R. E.*, **16**:521–535(1928).

Cady, W. G. "Piezoelectric Terminology," *Proc. I. R. E.*, **18**:2136–2142(1930).

Cady, W. G. "A Survey of Piezoelectricity," *A. J. Phys.* (formerly *Am. Phys. T.*), **6**:227–242 (1938).

*Cady, W. G. *Piezoelectricity*, New York: McGraw-Hill, 1946; reprinted in two volumes by Dover Publications, Inc., 1964.

*Frerking, Marvin E. *Crystal Oscillator Design and Temperature Compensation*, New York: Van Nostrand, 1970.

*Gerber, E. A. "Precision Control and Selection: A Bibliography," Proc. 33d *Annu. Symp. Freq. Contr.*, 1979, pp. 569–728. (This work lists over 3000 references covering the years 1968–1978.)

Heising, R. A. (ed.) *Quartz Crystals for Electrical Circuits.* New York: D. Van Nostrand, 1946.

Holland, R. and E. P. Eernisse. *Design of Resonant Piezoelectric Devices, Res. Monogr.*, no. 56, Cambridge: The M. I. T. Press, 1969.

**Index to Proc. 30th Annu. Symp. Freq. Contr.*, 1976, pp. 494–543. (This index lists all paper presented to Annual Symposium on Frequency Control during the preceding decade.)

Little, G. J. S. Eleven reports prepared for use within the British Post Office Engineering Departments, Post Office Research Station, London: Her Majesty's Stationery Office, 1957.

Mason, W. P. *Piezoelectric Crystals and their Applications to Ultrasonics.* New York: D. Van Nostrand, 1950.

*Mason, W. P. *J. Acoust. Soc. Amer.,* **41**(4): 907-912 (1967). (*A complete list of the publications of W. P. Mason.*)

Prichard, A. C. and M. Bernstein. "Crystal Impedance Meters," *Electronics,* May 1953.

Shockley, W., D. R. Curran, and D. J. Koneval. "Trapped Energy Modes in Quartz Crystal Filters," *J. Acoust. Soc, Amer.,* **41**:981–993 (1967).

Van Dyke, K. S. The Piezoelectric Resonator and its Equivalent Network. *Proc. I. R. E.,* **16**:742-764 (1928).

*Vigoreux, P. *Quartz Oscillators and Their Applications.* London: His Majesty's Stationery Office, 1939.

*Vigoreux, P. and C. E. Booth. *Quartz Vibrators and Their Applications.* London: His Majesty's Stationery Office, 1950.

INDEX

INDEX